9th
第九届金盘奖获奖作品集 空间类

THE 9TH KINPAN AWARD FILES SPACE DESIGN

品质/艺术/人居/价值
QUALITY, ART, HABITAT, VALUE

金盘地产传媒有限公司 策划
广州市唐艺文化传播有限公司 编著

最佳样板房空间
BEST SHOW FLAT SPACE

别墅户型
VILLA

大户型
LARGE HOUSE TYPE

中户型
MEDIUM HOUSE TYPE

小户型
SMALL HOUSE TYPE

最佳酒店空间
BEST HOTEL SPACE

度假酒店
RESORT HOTEL

精品酒店
BOUTIQUE HOTEL

商务型酒店
BUSINESS HOTEL

最佳售楼会所空间
BEST SALES CENTER SPACE

最佳商业空间
BEST COMMERCIAL SPACE

中国林业出版社
China Forestry Publishing House

品质 通过好的材质和施工技艺所呈现出来的优质感；

艺术 在开发层面有新的理念和模式；在产品设计层面有创新，契合时代发展需求；

人居 与环境完美结合，空间规划合理化、人性化，非常适宜居住；

价值 从产品本身讲，是一个时代的标杆；从商业角度讲，具有投资价值和升值空间。

QUALITY high quality presented by good materials and technique;

ART new concept and mode of development, and innovation of design according with the times;

HABITAT good combination with environment, and rational and human-orientated space, fit for living;

VALUE a benchmark of times from the view of product, and of investment value and appreciation space from the view of commerce.

前言

随着第九届金盘奖的圆满结束，2014年度的"中国好楼盘"榜单也已尘埃落定。作为金盘奖建筑类奖项的延伸，金盘奖空间类奖项同样立足于市场，着眼于设计，发起行业多方对话，力求为"中国好空间"寻找标杆项目。

金盘奖空间类奖项评选同样以"公平、公正、公开"为原则，经过严格筛选与权威论证，从全国300多个优秀项目中选出77个入围年度"最佳样板房、最佳商业空间、最佳酒店空间、最佳售楼会所"，最终选出20个项目作为"中国好空间"的年度标杆，让金盘奖所提倡的"品质、艺术、人居、价值"得到空间设计层面的解读，也让"中国好楼盘"的概念延伸至整个地产开发链中，进而能够更全面地总结地产设计方向与开发趋势，更深层次地探讨当代建筑、空间与人之间相互依存的关系。

作为第九届金盘奖空间类奖项的获奖作品集，本书在综合类作品集的编排风格基础上，根据空间项目的特点进行了相应调整，以实景图、设计图和设计说明为主要内容，用多元手法剖析优秀空间案例的设计理念、空间布局、材质应用、细部节点等，并附带现场嘉宾点评、设计师简介等内容，另外，书的编排以空间设计为核心，却并不局限在空间设计的范畴，涵盖建筑、景观实景图与楼盘介绍，让读者能够更全面地了解案例，更立体地感受一个好的空间所传达的内涵。

金盘奖空间类奖项关注的不仅仅是空间设计本身，更是把空间设计放在房地产开发的大链条中去讨论。在评选的过程中和这一本作品集当中，我们力图搭建起设计师、开发商与消费者之间的桥梁，让设计产生的价值更贴近市场需求，更符合时代潮流，进而孕育出更美好的生活。

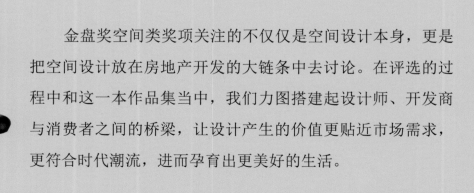

As the 9th Kinpan Award came to a successful conclusion, the "China's best property" list in 2014 was turned out. The Kinpan Award Space category, after Kinpan Award Architecture category, bases on market, focuses on design and initiates intra-industry multilateral talks to seek benchmark projects for "China's best space".

The Kinpan Award Space regards just, fair and open as evaluation principles. Having experienced strict selection and authoritative appraisal, 77 projects were elected among over 300 nationwide into four types: the Best Show Flat Space, the Best Commercial Space, the Best Hotel Space and the Best Sales Center Space; Ultimately, 20 projects win the honorary title of "The Benchmark of 'China's best space' 2014". The 9th Kinpan Award Files Space Design is conductive to deeper understand the "quality, art, habitat and value" evaluation purpose, meanwhile it brings "China's best property" concept into the whole developing links of real estate industry, and further to combine the design direction and development trend as well as to make a deeper discussion of the interdependent relations of architecture, space and human beings.

The 9th Kinpan Award Files Space Design inherits the compiling style of its prequel comprehensive category and adjusts with the features of space project. Real scene picture, design drawing and design specification play a dominant role in the book. In the same time it adopts multivariate technique to analyze design ideas, spatial layout, materials and details to display various spaces, and honored guests comments and designer's profiles are added as well to enrich contents. This book mainly concerns space design, while it break out the space scope to embrace architecture, real scene picture and property introduction so that readers can know more about every project and have an all-round view of the connotation of those preeminent spaces.

The Kinpan Award space category not only emphasizes space design, but takes the space design into real estate development links for discussion. During the Kinpan Award appraisal process and in this book likewise, we contribute to build a communication bridge for designers, developers and consumers, making the design value get close to market requirement and align with the times, and eventually breed a prosperous life.

人居空间：生活与家

金盘奖是立足于房地产领域且做得很精、很专业的一个设计奖项，该奖项从"品质、艺术、人居、价值"四个核心价值观上去研究及传播房地产开发与设计的相关资讯，并得到行业众多专家人士、设计师和开发商的认同与参与。每年的金盘奖评奖都秉着公平、公开、公正的原则，这个奖项对房地产乃至整个行业有着巨大的影响，对项目品质的提升也起到了毋庸置疑的推动作用。

在这个奖项中，我们强调房地产市场及产品设计的关联性，重视房地产住宅产品的设计功能性、系统性及可持续性，因为它是人们居住的空间。任何一个好项目，如果脱离了设计，都不可能更完美地呈现，我们强调设计，本身就是关注"人居"这个主题目标。

2014至2015年，是房地产市场极不确定的两年，房地产地块价格从高位开始下行，甚至部分地块流拍。同时政府对房地产市场起着微妙的影响，房产交易量也从巅峰下滑至新低；市场低迷，越来越显现出地产产品的优劣。在同等的促销条件下，品质较高的产品在市场中显现出较强的竞争力，一直处于卖方产品格局的状况有望在逆市中得以改善。越来越多的发展商在前期规划、建筑设计、景观设计及室内装修设计、施工质量、成本控制、产品营销等领域加大深化力度，力求以优势产品改变市场状况。

现今的很多地产产品过于注重高大上的形象，更着力于产品的形式及风格，甚至赋予项目艺术及文化范畴的属性；而当代艺术也对处于资本运营热点的房地产业表达了充分的热情，似乎追求艺术与高尚生活成了品味及身份的代名词。艺术介入人居空间并以此为行销手段，提升人居的文化形象及价值，在人居产品中体现了系统及专业化的美学经验，市场中大量出现所谓"新东方主义风格、现代风格、南加州风格"等人居产品。美学被过度的经验化，这个观点并不是否定美学的影响力，也并非质疑商业与艺术文化的共生环境，可是当美学被过分的操作和利用时，人居产品也就失去了作为人生存环境的初衷和内涵。其实我们应该更多地把精力用于思考人居层面的东西，思考怎样让空间照顾人的生活。日本住宅的空间设计虽然朴实，但胜在实用，人们居住在这样的空间里会觉得被空间所照顾，这应该是值得我们学习并且努力的方向。从人性化的居住角度来讲，我们的空间有太多的细节可以完善。例如，入口的空间虽小，但需要承载的功能很多：它要阻止外面的尘土被带进屋内；人们要在这里换鞋、放包；往深一层里说，这也是我们从社会回归家庭之后心情转化、行为转换的一个空间，可以在这里放下很多应该放下的东西。但在现在的很多产品中，门厅就只是一个窄窄的通道，不能满足实用功能以及用户的心理需求。

居住空间，从企业角度讲是产品，从居住者的角度则是具有情感归宿的"家"。从后者来讲，人居空间更应"产品化、工业化"，最好打造一个可以变化、生长的空间，而不只是一个固定化的空间。设计者应尽量在提供完善的厨卫及储藏等设施配套标准的前提下，让业主可以进行一些个性需求上的改造，自行去完善更多的功能，让业主的生活因此而更加便利和温馨，让家变得可以生长。除此之外，中国人还有家的情结，里面包含了很多情感因素。一个房子，一个家，里面住着爸爸、妈妈、小孩，可能还有爷爷、奶奶，怎样既保留家庭中的人情味儿，又让每个成员各得其所，是设计需要解决的首要问题。

除了居住空间，在商业空间、酒店空间、售楼会所等这些专业地产的板块，我希望金盘奖可以在房地产设计的细分领域做得更加专业，多抛出更多对行业对社会积极的观点。对于楼盘和空间的报道，多呈现发展商用功且未被人们关注的专业领域，也可以从消费者的体验角度，或者邀请专业设计师具体阐释开发商人性化的精细化设计。

第九届金盘奖空间作品集无论是对空间还是对楼盘的展示都远远超出了图片本身，在专业和技术上较有深度。项目展示擅于挖掘发展商产品的亮点，完美呈现吻合市场的卖点，对整个项目的设计理念、人性化的关注等进行了更深入地探索和阐述，承载了更多的东西，具有很强的专业性。我期待第十届金盘奖空间作品集更具学术性和可研性，向读者呈现更丰富的空间素材，引发必要的行业思考。

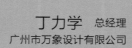 总经理
广州市万象设计有限公司

Living Space: Life at Home

Kinpan Award, basing on real estate industry, is a professional property design award, which regards quality, art, habitat and value as core values, and relies on these values to research the development and designing in real estate. The award has gained recognition and supporting from the industry experts, designers and developers. Because every Kinpan Award strictly abides by just, fair and open as evaluation principles, it has imposed huge influence on all over the real estate industry.

The Kinpan Award regards highly of the relation between real estate market and product design, and it also cares residence function, systematicness and sustainability since it is our living space. No excellent project exists without design, and the reason why design becomes the focus is that we care about human habitation.

The real estate market has been undulating in 2014 and 2015. The plot price has descended sharply, some plots auction even failed. Government anti-corruption measures play a subtle effect on real estate market in the process, which affects the property transaction volume to slump to a fresh low. Under such a background, the property quality appears to be more pivotal. Under the same conditions of sales promotion, higher quality properties show stronger competitiveness in the market, meanwhile seller controlling market situation is improved. More developers start to concern planning, architecture design, landscape design, interior decoration, construction quality, cost control and product marketing so as to adapt the market.

Nowadays, too much attention is paid on imposing image, property form and style, or even draws artistic and cultural features in. Modern art shows its enthusiasm on real estate, which becomes the focus of capital operation. Art and noble life seem to identify taste and status. Art is introduced in habitation space and acts as a marketing method to elevate habitat image and value, reflecting systematic and professional aesthetics experiences. Aesthetics is utilized empirically, which does not deny the influence of aesthetics or the coexisting environment of business and artistic culture. However, when aesthetics is overdone, habitation property loses its original intension and connotation as living carrier. In fact, we should transfer our insight to habitation and how to make a better space for inhabitants. Although Japanese living space design is plain and ungarnished, they care practicality. People live in such a place would feel being taken care of by the space, which is a correct direction we should learn and exert effort. From the aspect of human-orientation, there are many details we should improve in our space design. For instance, entrance is a small space, but it bears many functions: preventing dust in, changing shoes, unloading belongings, etc. Into a deeper meaning, it is an exchange space where we come home from outside. Here, we adjust our emotion and temporarily forget those complications. However in China, the fact is that the entrance is just a narrow vestibule without humanistic care.

For real estate companies, living space is a product while it is a home for every inhabitant. Because they need a living space that can be altered and developed instead of a fixed space. Residents can change their living space according to their own ideas. As designers we should lower decoration requirements while improve basic facilities so that customers can make simple decoration alteration and complement functions at ease. Beyond that, every Chinese regards home as a love knot which contains so much affection in. In a house, there is a papa, a mom, kids, and maybe a grandpa and a grandma living together. So how to keep the human touch while guarantee a proper private space for every member is the prerequisite of an excellent design.

Except living space, we hope the Kinpan Award could do more professional researches on sub-segments, such as commercial space, hotel space and sales center space. Meanwhile, it could introduce design details developers may neglect, or invite professional designers to interpret human-oriented design details.

The 9th Kinpan Award File Space Design introduced properties and space design. This is not just a display for architectural pictures. The book made deeper research in professional angles and technical aspects. It excavated highlights of every project and found their selling points matching the real estate market. Meanwhile it made intensive study on design concept, human-orientation, etc. I have an expectation that the coming 10th Kinpan Award Files Space Design could do more in academic and sci-tech aspects to bring readers rich space design ideas, and let them feel that they are exposing in those real scenes when they read the book.

Ding Lixue *General Manager*
Visov Design Co., Ltd.

嘉宾点评

王 冠
Wang Guan

最佳售楼处的特点有：第一，比较彰显地标性；第二，有一个前后功能的转换；第三，与建筑设计、周边环境和谐统一。

The best sales center has following features: apparent landmark location, function alternation and harmonious unification with architectural design and surrounding environment.

何永明
He Yongming

小户型需要用一些很艳丽的色彩来做样板房，也是属于"命题作文"。

Small house type needs showy colors to decorate show flat, so the design is like writing "a proposition essay".

张 宁
Zhang Ning

商业广场要突破模式化，有想法、敢创新，打造一个有价值感的商业空间。

Commercial plaza should break through modeling design by new ideas and innovation so as to create a valuable commercial space.

陈 颖
Chen Ying

加法太多的商业空间项目,总有一些想说话的冲动,反而会减弱本来空间的品质。

If add too much to a commercial space project, just like there is always an impulse to talk, which would scarify the original quality on the contrary.

丁力学
Ding Lixue

售楼部设计不应仅停留在做纯粹的设计上,更应该让业主体会到整个小区的文化氛围,消费群定位等。

Sales center design should no longer stay on pure designing, but strive to bring the community atmosphere to clients and clear out consumer positioning.

吴 滨
Wu Bin

小型的有意思的度假酒店将来在中国的市场需求上会有一个急速成长。

Small scale and enjoyable resort hotels will obtain a increasing growth in China.

韩 松
Han Song

主力户型的评判标准主要依靠视觉上的可辨识度。

Judgment criteria on an anchor house type depend on visual discrimination.

洪德成
Hong Decheng

做设计要勇于坚持自己作品的原汁原味。

Making design should insist on the original taste and style of our own works.

何宗宪
He Zongxian

优秀的样板房首先要满足市场的需求

A preeminent show flat should meet the demands of market first.

年度最佳样板房空间
The Best Show Flat Space 2014

样板房的品质体现在好的品味和好的材质以及好的施工工艺所营造的整体优越感；艺术层面，无论是硬装还是软装上，整体营造一种家居式的情境体验，具有引领性；人居方面，人性化的空间布局和设计，营造出舒适的生活体验；在价值的角度，产品既是一个优秀的设计，又让消费者有购买的欲望。

The quality of a show flat embodies in the overall superiority feeling created by good taste, wonderful materials and superior construction crafts; in the aspect of art, to create a household-like experience through hard and soft furnishing is a steering design; the humanized space layout and design bring comfortable habitat experience; and the value of a show flat lies in that it not only is a preeminent design but can stimulate purchasing desire.

别墅户型 VILLA

苏州仁恒棠北浅山别墅E户型样板房
Tang Island Type E Showroom, Suzhou — **18**
——高贵品位融于清净空间，时尚气息展露于细节之处
—Infuses noble taste into peaceful and tranquil space; Exuberate fashionable flair by delicate details

苏州新鸿基湖滨四季G户型样板房
Lake Genève G Show Flat, Suzhou — **28**
——现代简约的空间肌理中点缀丰富的精致细节，米色与深咖色带来平和愉悦心境
—Modern and concise space interspersing with exquisite details, beige and dark-coffee colors bringing peaceful and pleasant mood

北京万通天竺新新家园样板间
Legacy Homes Vantone Casa Villa, Beijing — **36**
——空间通透，功能布局合理，少量东方元素起到了画龙点睛的作用
—The space is transparent in a reasonable function layout, and a small amount of oriental elements bring out the crucial points

大户型 LARGE HOUSE TYPE

成都中德英伦联邦A区12#顶楼复式
British Ville A-12 Duplex Penthouse, Chengdu — **50**
——解构主义设计风格，充满前卫感和视觉冲击力
—Deconstruction design style, full of avant-garde sense and visual impact

北京万科北河沿甲77号样板房
Vanke Beiheyan Palace 77 Show Flat, Beijing — **62**
——气势磅礴，空间区隔得当
—Great momentum, with appropriate space partition

宁波财富中心示范单位
Financial Center Show Flat, Ningbo — 72
——扇形空间形态、合理的灯光照明、流动的线条设计形成了极富艺术魅力的互动空间
—Circular sector space, rational lighting and fluid lines comprise a charming interactive space

中户型 MEDIUM HOUSE TYPE

佛山怡翠宏璟样板间
Emerald Collection Show Flat, Foshan — 86
——现代主义的建筑形体理念在空间内的完美应用
—Modernism architectural form concept perfectly applied in space

中山时代倾城四期04户型样板间
Times King City Phase IV 04 Show Flat, Zhongshan — 96
——传统与现代、雅致与奢华并存的居住空间
—A coexistence of tradition and modernity, elegance and luxury living space

济南建邦原香溪谷二期D5户型样板间
Toscana Holiday Phase II D5 Show Flat, Jinan — 106
——将样板间展示与接地气的生活化场景相结合，呈现别致现代生活
—The display of the show flat combines with life scenes, presenting a unique contemporary life

小户型 SMALL HOUSE TYPE

佛山保利西雅图8栋A3样板房
Poly Seattle 8-A3 Show Flat, Foshan — 118
——直线、直角与三原色的空间组合，向抽象艺术家蒙德里安致敬
—A space of straight lines, right angles and three-primary colors, it pays our respects to abstract artist Mondrian

年度最佳酒店空间
The Best Hotel Space 2014

酒店的品质需要优质选材与精湛的工艺，更需要好的设计与规划，以及提供的有形与无形的服务；通过科学与艺术的结合获得酒店空间的灵魂与生命，设计有创新并能给人带来愉悦与舒适，都属于艺术层面；人居方面，合理的空间布局、齐全的功能配套、充满人性化的设计，处处体现出酒店的细心与用心；能让消费者有再次想住的欲望和宾至如归的感觉，就是酒店的价值所在。

The quality of a hotel needs refined materials and skilled craftsmanship, while splendid design, planning and tangible and intangible services are more crucial; the soul and life of a hotel are created by the combination of science and art, and an innovative design can bring pleasure and comfort; as for habitat, rational spatial layout, complete supporting facilities and humanized design express carefulness and attentiveness; to induce customers to reside again and have a feeling that it is a home away from home is the value of a hotel.

度假酒店 RESORT HOTEL

厦门乐雅无垠酒店
Hotel WIND, Xiamen — 130
——一个现代村落式的度假酒店，可沟通的居住氛围
—A village-pattern resort hotel, a communicative residing atmosphere

精品酒店 BOUTIQUE HOTEL

成都钓鱼台精品酒店
The Diaoyutai Boutique, Chengdu — 140
——宽窄巷子里的精致下榻，最具中国元素的别样风情
—An exquisite hotel in Kuan Alley and Zhai Alley, special amorous feelings of Chinese elements

商务型酒店 BUSINESS HOTEL

深圳四季酒店
Four Seasons Hotel, Shenzhen — 150
——现代艺术与传统元素完美融合，缔造典雅艺术空间
—A perfect integration of modern art and traditional elements, creating an elegant artistic space

曲阜香格里拉大酒店
Shangri-La Hotel, Qufu — 160
——新中式古典空间格调，承载浓郁的儒家情怀
—New Chinese classic space, bearing profound Confucian feelings

年度最佳售楼会所空间
The Best Sales Center Space 2014

售楼会所的品质体现在能够让客户对项目的品味以及档次产生最直观的感受，为楼盘代言，充当一种信息沟通使者的角色；艺术层面，强烈的视觉和示范作用，用艺术的方式营造"代入感"；人居方面，成熟的功能布局和配置，使顾客拥有良好的体验氛围；价值上则体现在产品的完整性、商业模式以及功能均经得起推敲，示范性设计能够刺激消费欲望。

The quality of a sales center lies in its intuitive feelings on taste and level, and it translates itself as an information communicator; intense visual impacts and demonstration bring people in the design in an art way; as for habitat, mature functional layout offers good experience ambiance; the value exists in the liable product completeness, commercial mode and functions, and its demonstration design stimulates purchasing desire.

黄山悦榕庄度假别墅区展示中心
Banyan Tree Resort Villa Show Center, Huangshan — 174
——清、雅、朴、华，古代徽派美学融入现代生活
—Feature in clear, refined, plain and unadorned, ancient Hui-style aesthetics blend in contemporary life

海南清澜半岛示范区
PE ninsula Demonstration Zone, Hainan   186
——尊重自然，崇尚人文的"木屋"
—A wooden house, respecting nature and admiring humanity

广州萝岗绿地中央广场售楼处
Guangzhou Greenland Central Plaza Sales Center   200
——独特的空间一体化设计、错层式功能布置，非常具有时代代表性
—Unique space integrated design and split-level function layout boast representativeness of the times

年度最佳商业空间
The Best Commercial Space 2014

商业空间的品质需要好的材质以及好的施工工艺，更需要在符合市场需求的条件下做出创新，塑造有价值感的空间；艺术方面，在体验的情境设计上有创新和引领性，营造出生活场所感；人居层面需要好的空间规划、业态组合和商业人流动线，整体营造一个好的购物体验，让顾客得到物质和精神上的双重享受；在价值层面不再仅仅局限于设计，意在打造有主题、有价值的空间。

The quality of a commercial space needs refined materials and skilled craftsmanship, and it also needs innovation under appropriate market conditions to create valuable space; use art technology to design innovative and steering experience scenario and life arena feelings; as for habitat, good spatial planning, formats combination and commercial pedestrian circulation bring user-friendly shopping experiences and meet material and spiritual double enjoyment; the value no longer limits in design, but in its themes and valuable spaces.

湖州东吴银泰城
Dongwu Intime City, Huzhou   212
——立足于小城市的消费心理需求，丰富美陈装置营造活泼商业氛围
—Basing on consumer psychology in small city, rich art decoration creates a lively commercial atmosphere

郑州锦艺城购物中心
Jinyi City Shopping Mall, Zhengzhou   222
——简洁设计打造华丽视觉效果，大气空间释放流畅时尚律动
—Concise design creates luxury visual effect, graceful space diffuses fluent and vogue rhythm

大连高新万达广场室内步行街
Wanda Plaza Indoor Pedestrian Street, Dalian   232
——多种设计手法穿插，细节把控合理，树立了A级店内装品质的新标杆
—Multiple design methods and rational grasping details establish a new benchmark of grade-A interior decoration quality

品质

与建筑外形完美吻合的空间形态，好的材质以及好的施工工艺营造出具有优质感的品质项目；根据空间特色进行功能区域划分，配以最适合的照明与装饰，更能够大幅度提升空间品质感，而好品质样板房的主要着力点是挖掘楼盘的文化内涵和购房者的居住需求。

艺术

充满新意的、具有艺术气息的样板房，总是能讨巧地运用色彩。在艺术方面，样板房需要针对目标受众，在硬装和软装上整体营造一种情境式的体验，设计完美的视觉效果与艺术氛围，产生一定的引领作用。

人居

从人居的角度来讲，样板房整体的完成度和完整度较高，给人以舒适感，不必过于讲求奢华，搭配协调，再加上完美的空间流线，合理的空间布局，舒服的软装搭配，人性化的设计，仿若天成般地塑造出完美的样板房空间。

价值

一个优秀的样板房所展现出来的价值是既能把设计者的想法、理念展示给大家，很好地帮客户控制造价，还能让购房者产生多停留一下，多看一点的意愿，有购买和居住的欲望。

Quality

High profile projects rely on quality materials, superior construction crafts as well as spatial forms echoing architectural appearance. To further improve a space needs accurate functional divisions by space features and appropriate lighting and adornments. Great show flats will highlight the exploration of property cultural connotation and the habitat requirements from buyers.

Art

Novel and artistic show flats invariably exert colors. Show flats should create a scenario experience through hard and soft furnishing aiming at intended buyers, meanwhile, guarantee perfect visual effect and artistic ambiance so as to guide the market.

Habitat

Show flats are generally of high completeness and completion, throwing off comfortable sensation. A perfect show flat needs no luxury decoration, but appropriate arrangement, fluent spatial circulation, cozy soft furnishing and humanized design.

Value

The value of an outstanding show flat should express the ideas and concepts of its designers, meanwhile help clients control cost and absorb buyers stay for a longer while and raise purchasing desire or even living desire.

随着房地产市场的逐渐成熟以及购房者的理性置业，样板房不再仅仅作为购房者装修效果的参照实例，也日益成为住宅文化的一种表现，让人们看到前沿的设计潮流，活灵活现地把今后的生活勾画在面前。衡量一个样板房的品质优劣有很多判断标准，比如整个项目的平面规划，风格化的装饰面和软装布置等。金盘奖嘉宾认为，优秀的样板房须是满足市场需求的主流作品，在视觉上具有可辨识度。金盘奖主要从品质、艺术、人居、价值四个方面来进行评判，带来前所未有的创新。

With the gradual maturity of real estate market and in the wake of rational buyers, show flats are no longer just references for decoration, it is a performance of residential culture which guides people to see cutting-edge design trend and spreads out a vivid living picture in front of them. There are various criteria to measure the qualities of show flat, such as plane planning, veneer styles, soft furnishing arrangements. However, Kinpan Awards participants hold that excellent show flats should be mainstream works in the market and be highly recognizable, so the Kinpan Awards innovatively evaluate projects from quality, art, habitat and value.

年度最佳样板房空间
The Best Show Flat Space 2014

苏州仁恒棠北浅山别墅E户型样板房
Tang Island Type E Showroom, Suzhou

苏州新鸿基湖滨四季G户型样板房
Lake Genève G Show Flat, Suzhou

北京万通天竺新新家园样板间
Legacy Homes Vantone Casa Villa, Beijing

成都中德英伦联邦A区12#顶楼复式
British Ville A-12 Duplex Penthouse, Chengdu

北京万科北河沿甲77号样板房
Vanke Beiheyan Palace 77 Show Flat, Beijing

宁波财富中心示范单位
Financial Center Show Flat, Ningbo

佛山怡翠宏璟样板间
Emerald Collection Show Flat, Foshan

中山时代倾城四期04户型样板间
Times King City Phase IV 04 Show Flat, Zhongshan

济南建邦原香溪谷二期D5户型样板间
Toscana Holiday Phase II D5 Show Flat, Jinan

佛山保利西雅图8栋A3样板房
Poly Seatt le 8-A3 Show Flat, Foshan

年度最佳样板房空间（别墅户型）
THE BEST SHOW FLAT SPACE 2014 (VILLA)

苏州仁恒棠北浅山别墅E户型样板房
TANG ISLAND TYPE E SHOWROOM, SUZHOU

颁奖词
Award Words

高贵品位融于清净空间，时尚气息展露于细节之处
Infuses noble taste into peaceful and tranquil space; Exuberate fashionable flair by delicate details

开发商：苏州中辉房地产开发有限公司 **Developer:** Zhong Hui Real Estate Co., Ltd.
项目地址：江苏省苏州市吴中区通州路 **Location:** Tongzhou Road, Wuzhong District, Suzhou, Jiangsu
设计公司：梁志天设计师有限公司 **Design Company:** Steve Leung Designers Ltd.
面积：592平方米 **Site Area:** 592 m²
主要材料：白枫木饰面、墙纸、银灰洞大理石、雪花白大理石、白枫木三拼地板
Major Materials: White Maple Veneer, Wallpaper, Silver Grey Travertine, Snow White Marble, White Maple Tree Compound Floor
摄影师：陈中 **Photographer:** Chen Zhong
采编：陈惠慧 **Contributing Coordinator:** Chen Huihui

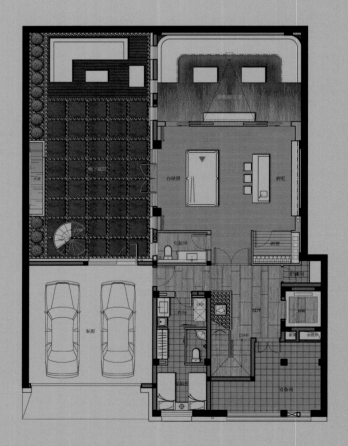

吴 滨 创始人
世尊设计集团

梁志天先生的作品一直表现得非常稳定、优秀，在这次评选的作品中一直比较领先。其整个房子的建筑平面规划、风格化的装饰面以及软装，整体水平都很高。

Wu Bin Founder
W+S Deco Group

Mr. Liang's works are always stable and excellent in design, and the selected work is in the lead as usual. Its Overall Floor Plan, stylish finishes and soft decoration are all in high performance.

嘉 宾 点 评
Honored Guest Comments

设计用简约的空间氛围表达一种大气优雅的生活态度，用精致的细节处理展现高贵的生活品位，用考究的选材与工艺缔造出时尚奢华的生活情趣。社交空间简洁明了，餐饮区域舒适宜人，卧房于低调之中点缀着品位非凡的艺术细节，再加上丰富的娱乐休闲空间，完美诠释当代新贵生活。

设计背景

仁恒棠北项目位于苏州工业园区，整个地块呈独立岛屿状，四面环水，规划有纯现代建筑风格的22栋滨湖独栋别墅，户型由400平方米至1 600平方米不等，建筑高度从东往西逐渐抬高，以满足每栋别墅对湖景的观赏需要。

设计思路：温馨雅致的空间品位

本案空间的主色调为米色，素净高雅，加上银灰洞石、进口枫木等高档建材，采用现代设计手法与优雅细腻的空间布局，并将秀丽宜人的天然湖景引入生活，为户主打造高级定制的尊贵品位空间。

Through concise spatial language and elaborated details, the space fully captures the discerning taste and living attitude of the households, displaying a luxury lifestyle of refinement and elegance, completed by selected quality materials and exquisite crafts. The social space is clean and forthright; the dining space is pleasant and comfortable; the bedroom is low-profile while dotted with extraordinary art details; coupled with plentiful entertainment space to perfectly annotate the newly rich life.

Designing Backgrounds

The Tang Island project situates in Suzhou Industrial Park, the plot is like an independent island with water all around it. The planning includes 22 waterfront single villas in contemporary style, and the house type is 400 m² to 1,600 m²; the height descents from west to east to satisfy the viewing desire for every villa.

Design Idea: Cozy and Elegant Space

The project takes plain, neat and elegant beige as dominant hue. It adopts high-grade building materials such as silver gray holey stones and imported maple wood. And under the contemporary design methods and in meticulous layout, the designers create a noble space for householders.

客厅、餐厅、书房

客厅布置优雅简约,以具肌理感的浅咖色扪布墙身配合银灰洞石地台,为大宅奠下雅致时尚的格调。淡灰色羊毛真丝地毯上,摆放着线条明快的浅灰色Giorgetti沙发及米黄色木质茶几,在水晶吊灯的映衬下散发出清新宜人的气息。

餐厅以优雅的Giorgetti扶手椅搭配长型餐桌,加上黑钛钢饰框的酒柜及开放式西厨,传达出简单优雅的生活态度。

同层的书房以浅色皮革及枫木为主材,搭配B&B Italia座椅及经典品牌灯具,营造舒适恬静的工作空间。

Living Room, Dining Room and Study Space

Light coffee cloth-upholstered wall and silver grey travertine anchor the room with neat touch of style and elegance. Light-gray Giorgetti sofa with clear lines and beige wooden table sit on the light-gray wool carpet, and with crystal pendent lamp lighting to emit a pleasant breath. In the dining room, elegant Giorgetti armchairs with long table, wine cabinet adorned with black titanium steel and open western kitchen express concise life attitude. In the same floor, the study space takes light color leather and maple wood as dominant materials, collocating with B&B Italia chairs and classic brand luminaire to offer a quiet working space.

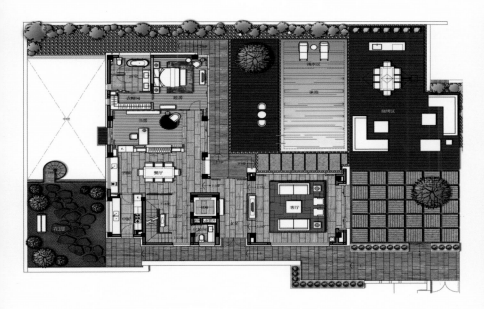

▼ 一层平面图
1F Plan

▼ 二层平面图
2F Plan

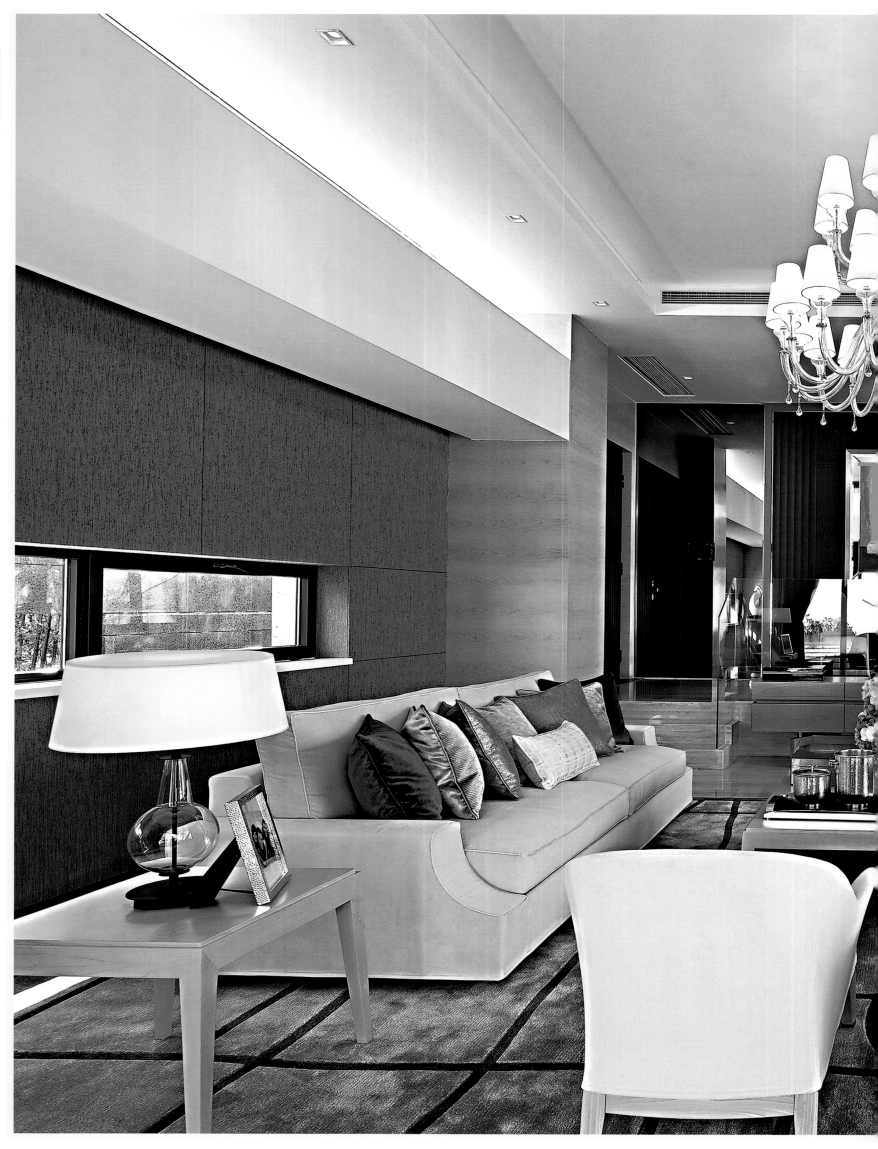

卧房空间：时尚奢华

本案采用双主卧布局。主人睡房一的背景墙以浅蓝色扪布饰面，配上茶镜造型框及华丽的水晶灯，加上Giorgetti系列家具和红酒吧，打造出高雅的空间氛围。踏着银灰洞石地台进入浴室，枫木饰面的独立浴缸及台盆柜搭配浴室镜面电视及桑拿蒸汽机，提供时尚奢华的卫浴体验。

主人睡房二以深色木饰面墙分隔卧室与工作间。一体化设计的大床搭配造型独特的黑色单人座椅及Flos落地灯，以鲜明的色彩对比凸显优雅的时尚韵味。浴室以雪花白大理石为主要石材，简洁的线条配合顶级设施，缔造舒适现代的卫浴空间。

娱乐派对空间

地下层的娱乐设施齐备，一室浅色进口木地板配衬台球桌及吧台，局部饰以深色皮革及灰镜，丰富整体的视觉效果。浅色扪布背景墙上错落有致地镶嵌着香槟金钢条，缀上一组别具玩味的挂墙摆件，洋溢一派盎然生机。推开两侧的木饰面移门，可将台球房、酒吧和影音室连为一体，方便举办各式派对及娱乐活动。影音室内，浅色错拼扪皮背景墙上设有宽大的屏幕，配合纯手工地毯、U型定制沙发及富有地方特色的装饰画，让生活更加丰富多彩。

Bedroom: Fashionable & Luxury

The villa has two master bedrooms. One of the master bedroom sets its bed against a background wall with light-blue cloth-upholstered cloth veneer, decorated with tawny glass frames and crystal lamps, Giorgetti furniture and wine bar counter. In the bathroom, there is a maple veneer independent bathtub, basin cabinet, mirror TV and sauna steamer offering luxury bathing experience.

The other master bedroom has a working room with dark wood veneer wall as partition. There is an integration king-size bed, a unique form black chair and Flos floor lamps. Its bathroom uses snow white marble as major materials with concise lines adornments and top facilities to offer a comfortable bathing space.

Entertainment Space

The basement is complete with entertainment facilities. Imported light wood floor is matched with billiard table and bar counter. Dark leather and gray mirror decorate at parts to enrich visual effect. The light cloth-upholstered background wall is inlaid with champagne steel bars and hung with an array of interesting furnishing articles, enhancing the vitality. Opening aside the wood veneer sliding doors, the billiard room, bar counter and video room are integrated as a whole, which is convenient for parties and entertainment activities. In the video room, there is a wide screen on the light staggered cloth-upholstered background wall, a pure manual carpet, a U-shape customized sofa and decorative pictures full of regional characteristics.

年度最佳样板房空间（别墅户型）
THE BEST SHOW FLAT SPACE 2014 (VILLA)

苏州新鸿基湖滨四季G户型样板房
LAKE GENÈVE G SHOW FLAT, SUZHOU

颁奖词
Award Words

现代简约的空间肌理中点缀丰富的精致细节，米色与深咖色带来平和愉悦心境

Modern and concise space interspersing with exquisite details, beige and dark-coffee colors bringing peaceful and pleasant mood

开发商： 新鸿基环贸广场房地产（苏州）有限公司　**Developer:** Sun Hung Kai Properties (Suzhou)
项目地址： 苏州园区金鸡湖大道888号　**Location:** No.888 Jinji Lake Avenue, Suzhou Industrial Park
设计公司： 梁志天设计师有限公司　**Design Company:** Steve Leung Designers Ltd.
承建商： 深圳市城建环艺装饰设计工程有限公司上海分公司　**Contractor:** Shenzhen Cheng Jian Huan Yi Design Co., Ltd. Shanghai Branch
面　积： 926 平方米　**Area:** 926 m²
墙身材料： 雀眼木饰面、墙纸、玲珑白玉大理石、雪花白大理石、新月亮古大理石、金摩卡大理石、白沙米黄大理石、圣罗兰大理石
Wall Body Materials: Bird Eye Maple Wood Veneer, Wallpaper, Jade Marble, Snow White Marble, New Moon Ancient Marble, Gold Moca Marble, Cream Pinta Marble, Saint Laurent Marble
摄影师： 陈中　**Photographer:** Chen Zhong
采编： 陈惠慧　**Contributing Coordinator:** Chen Huihui

王 冠 主持设计
矩阵纵横

梁老师的作品一眼就可以看到其使用的材料和施工水平均比较高，并利用讨巧的方法把设计师的想法完整地展示出来，又能满足市场的要求，非常厉害。

Wang Guan Chair Designer
Matrix Design Group

From the construction materials and skills we can know the work is accomplished by Mr. Liang. He is adept at completely expressing the designer's idea and meeting the needs of the market in the same time.

嘉 宾 点 评
Honored Guest Comments

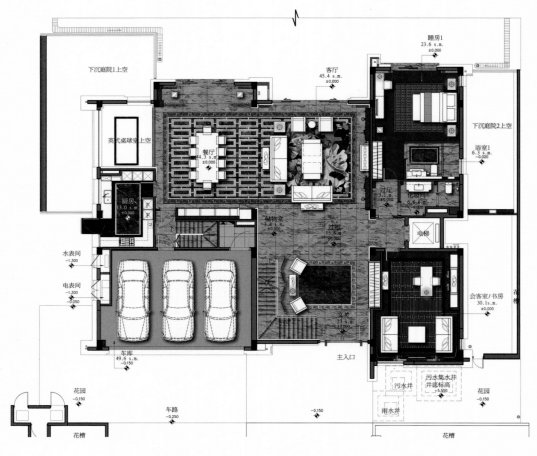

▼ 一层平面图
1F Plan

▼ 二层平面图
2F Plan

空间整体大气简洁，细节处又展现出设计的精致追求，流畅而奢华。米色与深咖色是本案空间的主色调，有鲜明而不冲突的明暗对比，理性之中蕴含着温暖柔和的意象，给人舒适宁静的空间体验。

设计背景

该样板房位于样板区的临湖位置，拥有超大湖景私家花园，大幅玻璃幕墙、双层挑空中庭、环绕式窗外露台及通透的室内空间让住户足不出户就能欣赏金鸡湖美景，将"湖中湖"的诗意景致延伸入室。

设计思路

设计师将项目临近花园和湖景的优势最大化，采用米色和深咖色为空间主色调，用现代简约的设计语汇表达中西方艺术元素，力求呈现品位非凡、情趣盎然的生活体验。

玄关、客厅及餐厅

一入玄关，一组饰有水晶条及香槟金镜钢框的巨型玻璃屏风映入眼帘，搭配具有中式韵味的家具及地花，令层高6米的空间倍显尊贵大气。

客厅的新月亮古石地台以雅士白石及黑白根石拼砌成中式地花，与白色欧式造型天花相映成趣。客厅背景墙以玲珑白玉拼花饰面，在水晶吊灯下尽展奢华气势。饰有银杏叶图案的地毯上摆放着美国JiunHo品牌家具，搭配色彩明快的饰品，为大厅增添几分温馨暖意。

餐厅的天花上，一盏波浪形水晶吊灯徐徐垂下，灯光洒落在深啡色雀眼木饰面餐桌及橙色扣布餐椅上，散发瑰丽和谐之感。厅侧设有精致现代的装饰柜，陈列各式收藏精品，为空间增添了高雅品位。

家庭厅及主人房

沿云石阶梯步至二层的家庭厅，啡色菊花图案地毯上摆放着优雅的弧形沙发组合，与中式造型电视墙相得益彰，营造温馨舒适的氛围。家庭厅后方是主人睡房，沉稳大气的紫色花纹地毯为房间注入了温馨自然的格调。超大比例白色大床，加上饰有香槟金镜钢的扣皮造型墙，散发矜贵奢华的时尚韵味。踏着新月亮古石地台进入主人浴室，可以看见大理石饰面的梳妆台、两侧的独立式发光白玉石洗手台、多功能按摩大浴缸，在一室雪花白的衬托下，倍显低调奢华。过厅另一边为独立衣帽间，简约优雅的扣皮陈列架、独立首饰柜、开放式衣橱，缀上精巧雅致的水晶吊灯，为户主打造尊贵时尚的生活空间。

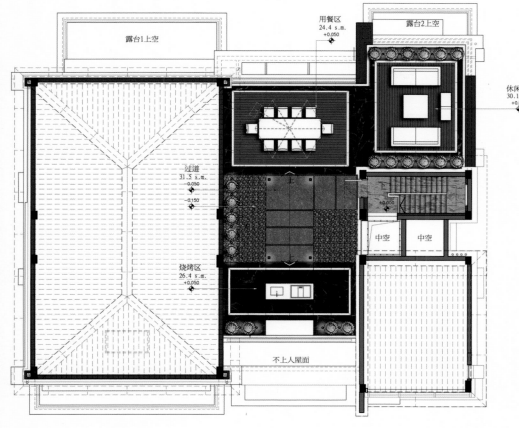

▼ 三层平面图
3F Plan

The overall space is grand and concise, while the details are exquisite. Beige and dark-coffee are the dominant hues, expressing a bright yet harmonious contrast. Warm and soft feeling consists in the rational state, offering a comfortable and tranquil space.

Designing Backgrounds
This show flat is beside the lake, owning super-size lake-side private garden. There are huge glass curtain walls, a two-floor-height atrium, a surround-type terrace outside of window and transparent inner space to allow hosts to appreciate the beauty of Jinji Lake.

Design Idea
The designer maximizes the advantages of garden and lake scenes and take beige and coffee colors as dominant hues. He also uses contemporary design methods to integrate Chinese and Western art elements, creating a conceited and noble living experience for householders.

Vestibule, Living Room and Dining Room
When one steps into the vestibule, a huge glass screen immediately jumps into sight, which is adorned with an array of crystal bars and a champagne-gold mirror steel frame. The furniture and the floor are in Chinese style, and the 6 m high space looks honorable and grand.

The new moon ancient marble floor platform in the living room and the floor in the ariston stone and black-white root stone patterns are echoing with the white European ceiling. The background wall in living room is decorated with white jade veneer, which looks luxury under the lighting of crystal lamps. American JiunHo furniture sits on ginkgo pattern carpet, added bright color ornaments, increasing warm and sweet breath to the living hall.

A wavy crystal pendant lamp hangs on the dining room ceiling, and the lighting projects on dark-coffee bird eye maple wood veneer dining-table and orange leather-upholstered dining-chairs, emitting a feeling of magnificent harmony. In an exquisite modern style sideboard, collection articles are displaying on, enhancing elegant taste.

Family Room and Master Bedroom
Marble stairs leads one to the family hall on 2nd floor, a set of arc sofas lying on chrysanthemum coffee pattern carpet complementing with Chinese style TV screen perfectly. At the rear of the family room is the master bedroom, calm and grand purple pattern carpet increasing cozy and natural flair. The large scale white bed and the champagne-gold mirror steel leather-upholstered walls emit luxury flavor. Walking on new moon ancient marble floor platform into master bathroom, a marble veneer dresser, independent glowing white jade wash basins at two sides and a multi-functional massage bathtub, foiled with snow white marble, look low-key and luxury. There is an independent cloakroom at the other side, with a concise leather-upholstered display stand, an independent jewelry cabinet, an open-style wardrobe, adorned with crystal pendent lamps, creating a noble and fashionable living space for householders.

年度最佳样板房空间（别墅户型）
THE BEST SHOW FLAT SPACE 2014 (VILLA)

北京万通天竺新新家园样板间
LEGACY HOMES VANTONE CASA VILLA, BEIJING

颁奖词
Award Words

空间通透，功能布局合理，少量东方元素起到了画龙点睛的作用
The space is transparent in a reasonable function layout, and a small amount of oriental elements bring out the crucial points

开发商：北京广厦富城置业有限公司　**Developer:** Beijing Guang Sha Fu Cheng Properties Co., Ltd.
项目地址：北京市顺义区天竺镇　**Location:** Tianzhu Town, Shunyi District, Beijing
设计公司：北京根尚国际空间设计有限公司　**Design Company:** GENSSUN
设计总监：王小根　**Design Director:** Wang Xiaogen
建筑设计：北京墨臣建筑设计事务所　**Architecture Design:** MòChén Architects and Designers
项目面积：365平方米　**Site Area:** 365 m²
竣工时间：2013年　**Completion Time:** 2013
主要材料：白色混油木饰面、手绘壁纸、铜条、拼花木地板、地毯等
Major Materials: White Oil-mix Wood Veneer, Hand-painted Wallpaper, Copper Bar, Parquet Wood floor, Carpet
采编：陈惠慧　**Contributing Coordinator:** Chen Huihui

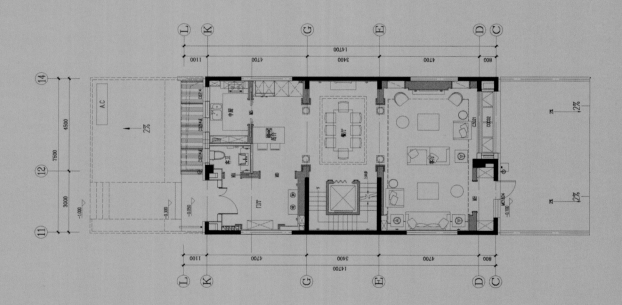

王 冠 主持设计
矩阵纵横

作品有新意，色彩运用也比较讨巧，能够把设计师的功力很好地体现出来。梁老师的作品我们一眼可以看到其使用的材料和施工水平，但是这个作品用讨巧的方法，既把设计师的想法展示了出来，又能满足市场的要求，还可以很好地帮客户控制造价，总之这个作品传统里面充满新意。

Wang Guan Chair Designer
Matrix Design Group

The project is innovative, and it does well in color collocation, from which we can know the designers' ability and skills. From the construction level and materials, the project designed by Mr. Liang is self-evident. The design not only expresses designers' idea but also satisfies market, well controlling cost for customers. In a word, the project is full of new conception.

丁力学 总经理
广州市万象设计有限公司

虽然是传统的欧式，但整个作品的完成度和完整度很高，而且在欧式的体系里面有一些创新，一些跟现代元素的结合点。我们从画面上很难看到一些比较明显的不协调的地方，它从大的空间关系和装修的细节上，在样板房方面把握得还是比较完整。

Ding Lixue General Manager
Visov Design Co., Ltd.

Although the project is in traditional European style, it has high degree of completion and integrity. It takes some modern elements in European architecture system, it is an innovation. From a picture perspective to observe the project, there is no inharmonious point, not only in space relation but in decoration details.

嘉 宾 点 评
Honored Guest Comments

设计师将自由、欢快的心情用最直接的手法呈现，让人在第一时间拥有最强烈的视觉体验。空间比较注重顺畅的动线与视觉的通透感觉，风格上采用法式新古典格调与现代抽象艺术元素相结合，并加入少量东方元素，在奢华中增添了些许大气与庄重。

设计背景

区域概况

项目位于北京市区东北方向，在五环路外，京承高速与机场高速之间，距刚刚投入使用的新国展中心仅1公里，距市中心约30公里。用地东接北京首都国际机场和规划中的国门商务区，南临温榆河及中央别墅区，周边还有多所国际学校和国家会计学院。地理区位非常优越，是一片极具开发潜力的热土。

建筑设计

本案的建筑设计渊源可以追溯到意大利托斯卡纳的红酒名镇蒙塔奇诺。其开发理念是蒙塔奇诺别墅式样的原版再现，也源自对独栋与联排别墅特色的精准把握，既融汇了联排别墅的精要，又因其顶楼区隔而具有独栋别墅的部分特征，故美该别墅名曰"独联体"。

The designers use direct methods to express free and cheerful mood, giving an intense visual experience when it is looked in the first time. The spaces lay emphasis on streaming circulation and transparent vision. They integrate French Neo-classic style and modern abstract art elements, adding a small amount of oriental elements to enrich space with grandeur and dignity.

Designing Backgrounds

Location

The project situates at northeast of Beijing, beside 5-ring road, between Beijing-Chengde Highway and Airport Expressway, just 1 km from China International Exhibition Center which is just put into use, and 30 km from downtown. The site is adjacent to Beijing Capital International Airport and planning Air CBD at east, and neighbors Wenyu River and Central Villa Zone. There are many international schools and the National Accounting Institute. It is a land of most development potential.

Architecture Design

The design can be traced back to Montalcino, a famous town of red wine in Tuscany, Italy. The design concept is to originally represent Montalcino villa pattern, which is also a precise grasp of the features of single villa and townhouse. It not only integrates the essence of townhouse, but also differentiates a part of features of single villa, and that is why the villa is called "single-united entity".

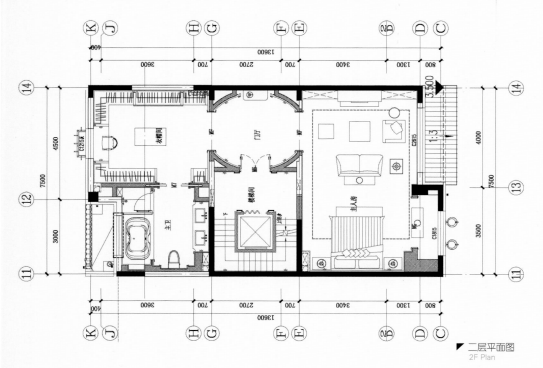

▼ 二层平面图
2F Plan

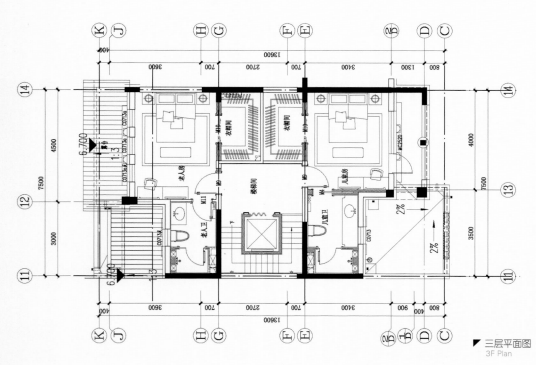

▼ 三层平面图
3F Plan

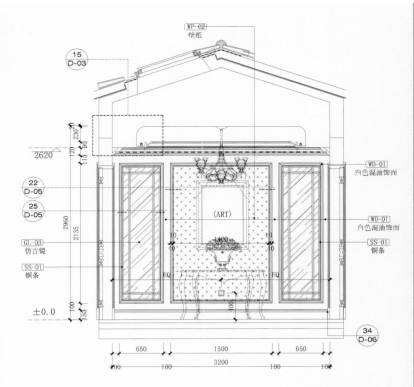

▶ 电梯厅立面图 1:30
Elevator elevation 1:30

功能布局

整个户型是一栋上下三层的别墅项目。一层为公共区，二层为主卧区，三层为老人房与儿童房，地下面积是休闲区，上下楼层设计了电梯来辅助行动。

设计风格

这是一个法式新古典格调与现代抽象艺术元素相结合，并伴有少量东方元素的空间。优雅的法式经典蓝、黄色丝质布艺与抽象地毯纹样、挂画、灯具相搭配，展现了精致、细腻兼具自由、奔放的空间格调。少量东方文化元素的运用，在奢华中增添了些许大气与庄重。

空间色彩

整体空间以蓝、黄色为主基调，使用中性的黑、白色中和空间色彩。入口玄关处的墙面使用浅色护墙板与做旧镜面来搭配。右侧的餐厅与客厅区域使用了4根白色"多里克式柱"使空间更加大气，白色柱体自然地融入空间。

Function Layout

This is a 3-floor house. The ground floor is for public activity, the second is bedrooms, the third is for the old and children and the basement is leisure area. The house has an elevator conducive to go up and down.

Design Style

It integrates French Neo-classic style and modern abstract art elements, adding a small amount of oriental elements to enrich space with grandeur and dignity. Elegant French classic blue and yellow silk fabrics, abstract pattern carpet, pictures and luminaires express an exquisite, delicate and free space.

Space Color

Blue and yellow are the main hues, with neutral black and white to neutralize space colors. The wall in vestibule uses light-color panel matching faded mirror; at right, in the dining room and living room stand 4 white "Doric Order", increasing the grandeur of the space, and the white pillars blend in space naturally.

室内设计

客厅

横向7.5米的客厅面积,设计两个会客功能区,使利用率得到充分发挥;并利用相同的靠枕、单人沙发,既保持了空间的独立性又不乏统一性。

主卧

整个二层通过玄关分为左右两个连贯的区域,右边为卧室区、阅读区,左边为步入式衣帽间和主卫间,使整个空间通透,功能布局安排合理。低纯度的蓝、黄色增加了空间的柔和与舒适,地毯的抽象图案是现代艺术的表现手法,而蓝色的使用使得现代元素和谐地融合在整体空间。

儿童房

儿童房格局特殊,是一个顶面高点为5米的竖形空间,为了避免较高的尺度造成的压抑感,设计以"降低线脚"和"隐形灯带"的处理手法,将空间分割为上下两段,视觉和感受上更加轻盈、空灵,空间变得舒适宜居。

老人房

老人房为咖色与乳白色的搭配,更加适合老人对安静、祥和居所的诉求,面料的光泽度增添了空间的优雅。

休闲区

具有休闲娱乐功能的地下面积,设计了健身区、红酒区。同样的"多立克式柱"与整体设计手法融会贯通,通透的格局也使在酒吧区品酒或者打桌球都拥有了开阔的视觉感受。

多立克柱式——是古典建筑的三种柱式中出现最早的一种（公元前7世纪），源于古希腊。特点是比较粗大雄壮，没有柱础，柱身有20条凹槽，柱头没有装饰，多立克柱又被称为男性柱。最早的高度与直径比为6:1，后来改至7:1。

Doric Order — is one of the earliest columns used in classic architecture (the 7th century B.C.), originated from Ancient Greek. It is very bulky without plinth, and it has 20 grooves on the body. Its chapiter has no adornments. The Doric Order is called masculinity column. The original ratio of height versus diameter is 6:1, then switched to 7:1.

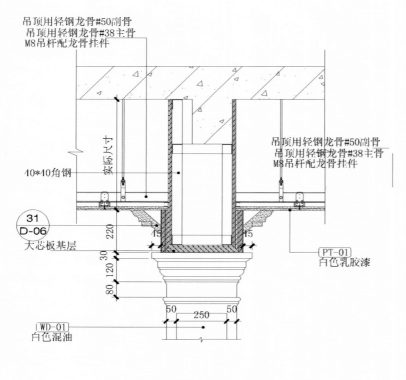

▶ 餐厅、客厅吊顶节点大样图 1:15
Dining Room and Living Room Suspended Ceiling 1:15

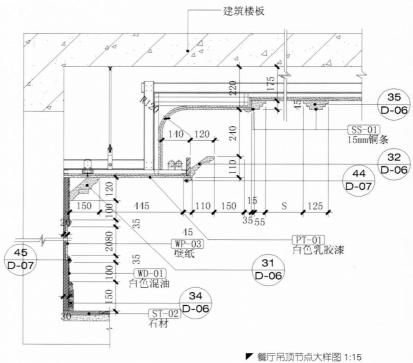

▶ 餐厅吊顶节点大样图 1:15
Dining Room Suspended Ceiling 1:15

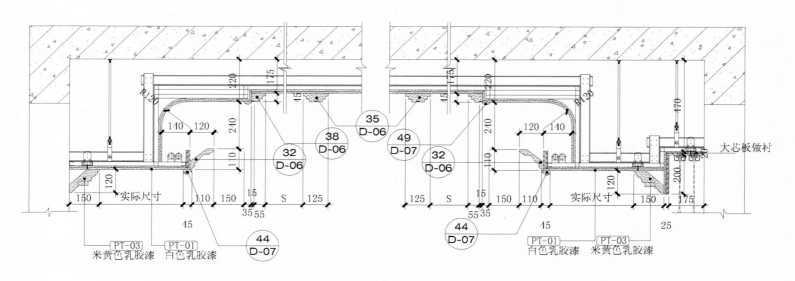

▶ 客厅吊顶节点大样图 1:15
Living Room Suspended Ceiling 1:15

▶ 客厅窗帘盒吊顶节点大样图 1:15
Living Room Curtain Box 1:15

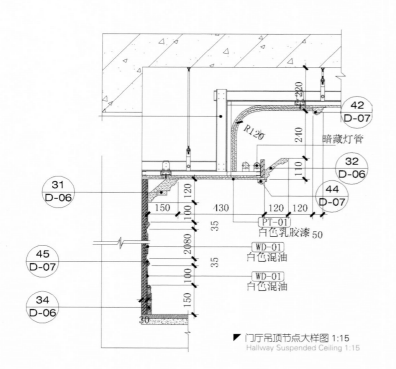

▶ 门厅吊顶节点大样图 1:15
Hallway Suspended Ceiling 1:15

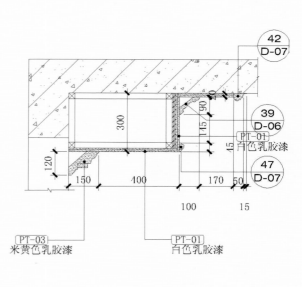

▶ 红酒吧吊顶节点大样图 1:15
Wine Bar Suspended Ceiling 1:15

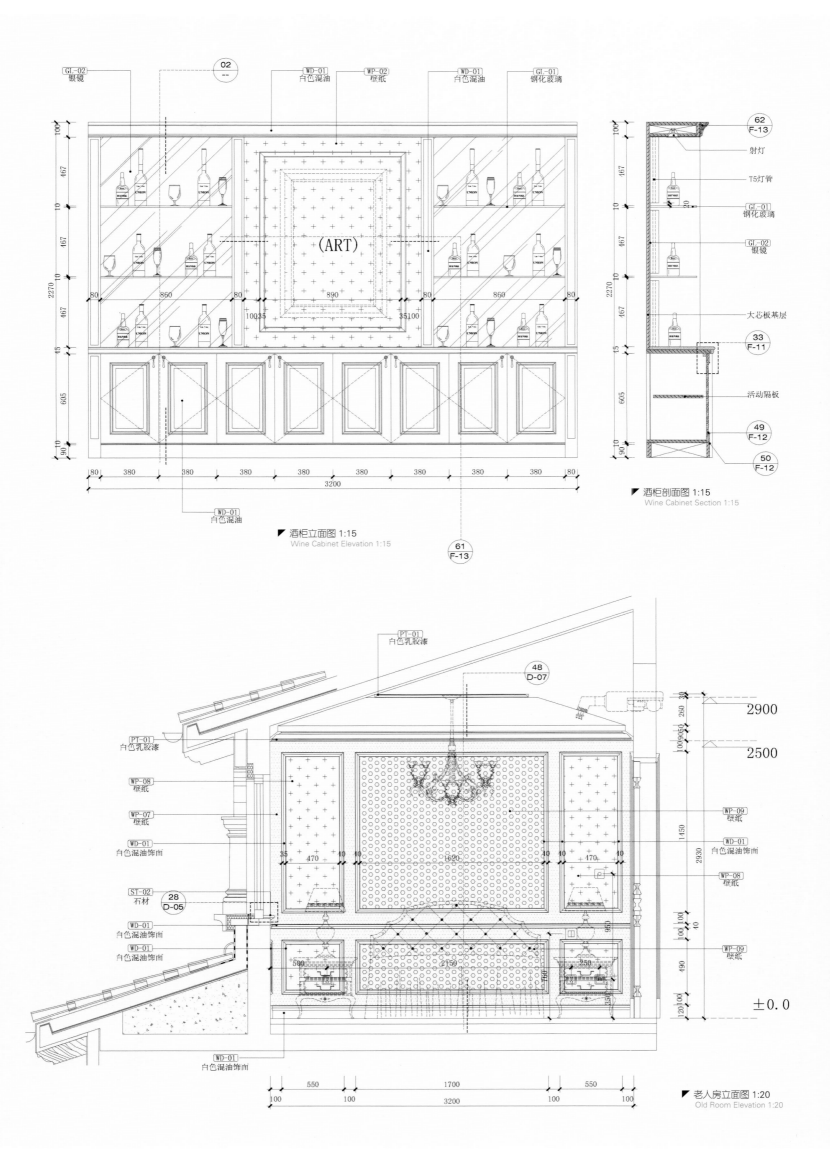

Interior Design

Living Room
The living room is of 7.5 m width, so it sets with two visitor receiving areas, which enhances area utilization ratio. The same cushions and armchairs keep the independence and unification simultaneously.

Master Bedroom
The 2nd floor is divided into right and left connected areas by a vestibule. Bedrooms and a reading area are at right side; and a cloakroom and the master bathroom are at left side. The overall space looks transparent and in a reasonable layout. Low-purity blue and yellow make the space soft and comfortable; abstract patterns in carpet reflect modern artistic taste; and especially, the blue makes the modern elements blend in the space harmoniously.

Children's Room
The children's room is special in layout, due to it is of 5 m high. In order to reduce the repression, designers use "lowering architrave" and "invisible light" methods to divide the space into two sections, so it looks more lithe and comfortable.

Old Room
The old room takes coffee and cream as dominant hues to cater to the needs of quiet and peace. The glossiness of decoration materials increases an elegant breath.

Leisure Area
The basement sets a leisure area, including a fitness zone and a red wine bar. It adopts "Doric Order" to keep unity with the overall design style. The transparent layout makes people, no matter sitting in red wine bar or playing at billiard table, enjoy a broad vision.

王小根 | Wang Xiaogen

先后毕业于沈阳航空工业学院设计系，中国科学院研究生院国际工程项目管理专业，获得IPMP证书C级证书认证。2013北京大学酒店管理总裁班在读；中国十大样板房设计师，IAI亚太设计师联盟理事。

Designer Wang Xiaogen, graduated from Department of Designing in Shenyang Institute of Aeronautical Engineering, and Graduate School of Chinese Academy of Sciences majoring in International Engineering Project Management. He got IPMP C level certificate authentication, and studied at Beijing University Hotel Management President Class in 2013. Mr. Wang is a China's top ten model house designer and a member of IAI (Asia-pacific Designer Alliance).

所获荣誉 Awards & Honor

2008年度中国十大样板房设计师

China's Top Ten Model House Designer 2008

公司简介

根尚国际的"genssun"英文含义是追求阳光和力量的团队。设计理念：立足传统文化的沃土，结合当下时尚的现代生活，表达出自己刚毅而柔美的设计风格。

Design Company Profile

The Company name "GENSSUN" means a team to pursue sunlight and power. They hold the design concept: keeping a foothold on traditional culture, combining present fashionable life and delivering resolute yet feminine design style.

代表作品

天津万通华府样板间、北京8哩岛样板间、成都玲珑郡（巴黎玫瑰）、内蒙鄂尔多斯华府世家B1户型、东方纽蓝地·B4户型(香榭丽舍)、东方纽蓝地·E3户型（松石绿的诱惑）

Representative Works

Vantone Mansion Show Flat, Beautiful Natural Island Show Flat, Chengdu Ling Long County, Neimeng Art of Living B1 Show Flat, Dong Fang New Land·B4 House Type (Champs Elysees), Dong Fang New Land·E3 House Type (The Temptation of Turquoise)

▼ 北京8哩岛样板间
Beautiful Natural Island Show Flat

▼ 成都玲珑郡（巴黎玫瑰）
Chengdu Ling Long County

▼ 内蒙鄂尔多斯华府世家 B1 户型
Neimeng Art of Living B1 Show Flat

▼ 东方纽蓝地·E3户型（松石绿的诱惑）
Dong Fang New Land·E3 House Type (The Temptation of Turquoise)

年度最佳样板房空间（大户型）
THE BEST SHOW FLAT SPACE 2014 (LARGE HOUSE TYPE)

成都中德英伦联邦A区12#顶楼复式
BRITISH VILLE A-12 DUPLEX PENTHOUSE, CHENGDU

颁奖词
Award Words

解构主义设计风格，充满前卫感和视觉冲击力
Deconstruction design style, full of avant-garde sense and visual impact

设计单位： 柏舍设计（柏舍励创专属机构） **Design Company:** PERCEPT DESIGN(PERCEPTRON GROUP)
项目地点： 成都市武侯区 **Location:** Wuhou District, Chengdu
项目面积： 约660平方米 **Site Area:** around 660 m²
竣工时间： 2013年12月 **Completion Time:** December, 2013
主要材质： 皮革、不锈钢、工艺玻璃、石材等 **Major Materials:** Leather, Stainless Steel, Craft Glass, Stone
采编： 谭杰 **Contributing Coordinator:** Tan Jie

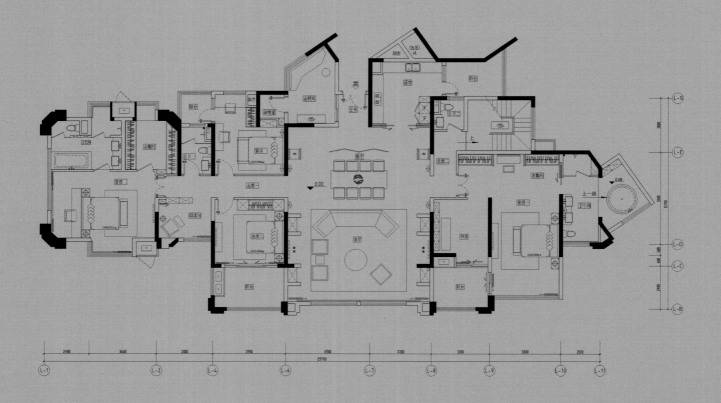

洪德成 创始人
DHA洪德成设计顾问（香港）有限公司

这个项目将近有660平方米，设计的整体气氛，如视觉效果、艺术气氛、空间感觉，都控制得比较统一，在评选的项目里面相对比较成熟，设计含量高，包括它的硬装和软装的搭配，还有公共空间都做得比较好。

Hong Decheng Founder
Dickson Hung Associates Design Consultants

This project is around 660 m², and it has a comprehensive plan—visual effects, artistic ambiance and spatial feeling with a sense of wholeness. It is relatively mature among the selection projects, and it did great in design, the collocation of hard and soft decorations as well as its public space.

嘉宾点评
Honored Guest Comments

本案中，美不再是抽象，而是体现在由空间、灯光、线条、质感共同营造出的意境中，同时也是设计师出于对人文和美学的尊重而呈现的艺术之美。解构主义的设计风格将整个空间用无规则的直线勾勒出来，配合后现代主义风格的基础材料，营造出强烈的视觉冲击力，充满前卫感和艺术感。

设计背景

区位特征

中德英伦联邦项目位于国际城南天府大道南段，地处天府新区核心区域，北临5平方公里的西部金融中心——金融城，东接规划为10平方公里的花园式新川创新科技园区，是国际城南首屈一指的超级大盘，也是城南未来20年最具发展潜力的新川板块旗舰项目。

规划理念

项目位于素有"天府之国"美誉的成都，故以英伦文化与生活气韵交融的古典风情为蓝本，精心打造涵盖住宅、高级会所、主题商业、双语幼稚园等多种业态的高端国际社区和专属于青年才俊的高尚英伦风情主题社区。

Beauty is no longer abstract in this project, but presented by artistic conception of space, lighting, lines and texture. Humanity and aesthetics are valued by designers. The deconstruction design style is applied to make irregular lines outlining space, complementing post-modern style materials, and binging strong visual impact, full of avant-garde sense.

Designing Backgrounds

Location

The British Ville locates at south of Tianfu Avenue in International City South where is the core area of Tianfu New District. The Financial City—western financial center of 5 km2 is at north, the planned Singapore-Sichuan Hi-tech Innovation Park of 10 km² is at east. The British Ville is a second to none super property in International City South, meanwhile, it is the most potential development project in Singapore-Sichuan Hi-tech Innovation Park over the next 20 years.

Planning Concept

The project is in Chengdu where is known as the "land of abundance". Based on British culture and classic expressions, it establishes a multi-formats high-end international community containing residence, club, commercial and bi-lingual kindergarten, and it is also a British style community for young elites.

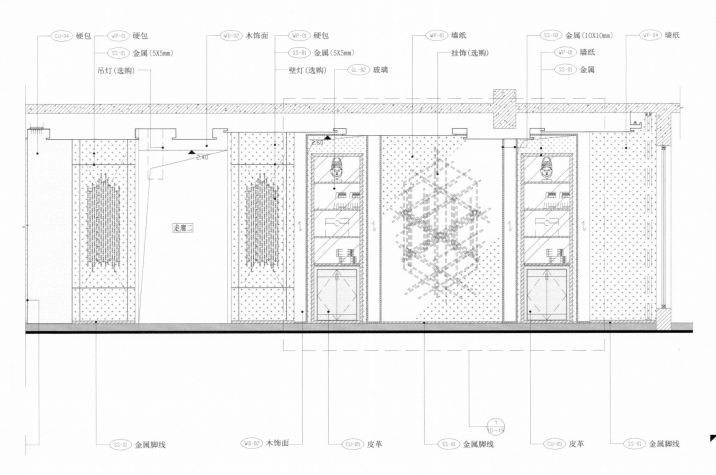

▼ 客厅立面图
Living Room Elevation

诠释生命的雕塑、
象征自由的面具

Sculpture:
interpretation of life,
Mask: symbol of
freedom

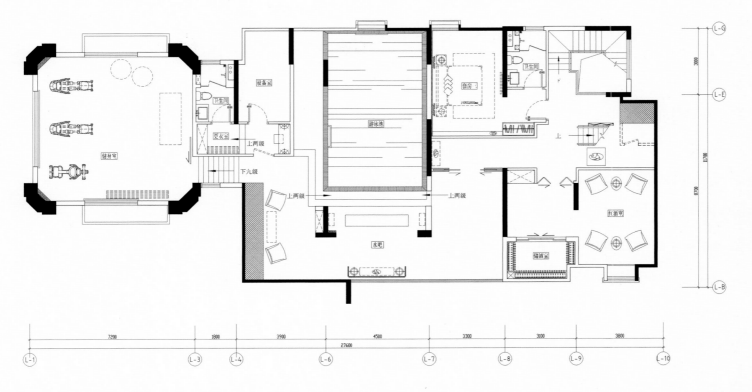

▼ 二层平面图
2F Plan

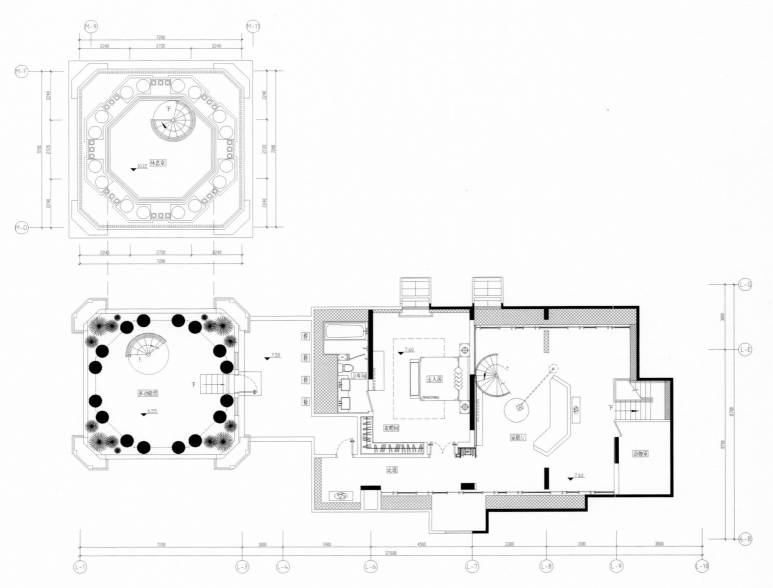

▼ 三层平面图
3F Plan

设计灵感

本案的设计灵感源于设计师对艺术与生活的理解，生命与力量的尊重，每一处每一寸每一方都似乎在谱写一些关于生命的篇章。

解构主义风格

设计师通过对空间的考量，采用解构主义设计风格，将整个空间用无规则的直线勾勒出来，强烈的前卫感和视觉冲击力，给人们带来全新的审美趣味和美感享受，让人印象深刻。除却强烈的艺术感，设计更是一场高级私人定制的大party，空间充满着法兰西的浪漫情怀，意大利的热情奔放，既有理性的流线也有感性的质感，更像一个国际化的舞台。

特色材质

为配合整个空间主题，设计师以后现代主义风格为基础对材料进行选择，如使用天花板的工艺玻璃对空间进行反射、折射，营造错位效果等，增强了整个空间的自由度和艺术感。

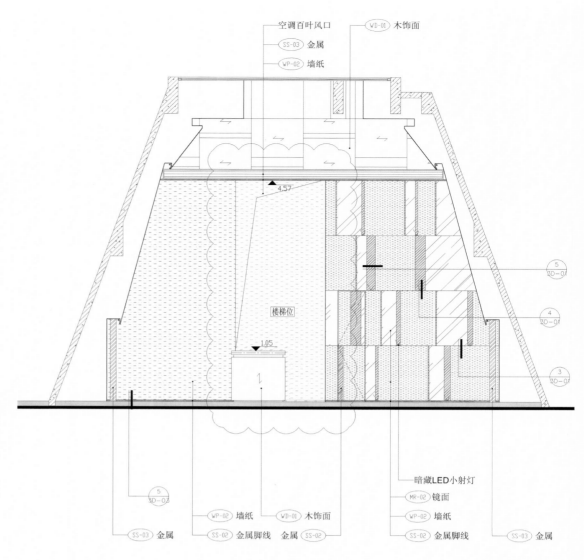

▶ 家庭厅立面图
Family Room Elevation

Design Inspiration
The design inspiration of this project comes from the designers' understanding of art and living and the respect for life and power; and every inch is the expression of pages of life.

Deconstruction Style
The deconstruction design style is applied to make irregular lines outlining space, binging strong avant-garde sense and visual impact, which flaunts its aesthetic taste and leaves profound impression. Besides the artistic feeling, the design creates an advanced personal tailor party, with France romantic flavor and Italian passion overflowing all the space. Rational fluid lines and nastic texture are balanced, making the space like an international stage.

Featured Material
In order to match the space themes, designers adopt post-modern style materials. The craft glass on ceiling has reflection, refraction and mismatching effects, strengthening the flexibility and artistic sense for the overall space.

年度最佳样板房空间 | THE BEST SHOW FLAT SPACE 2014

司徒曦 | Jason Situ

司徒曦，曾组织参与多项大中型室内设计项目，对大型别墅、示范单位及房地产会所设计有着独到的见解。擅于从原建筑固有特点及优劣势入手，优化其功能布局，熟练运用线条、色彩及材质使作品达到美感和实用完美结合。其曾获得多项国内外比赛大奖。

Jason Situ organizationally participated in many large and medium scales interior design projects, and has insightful views on large scale villa, show flat and real estate club. He is adept at utilizing the strengths and weaknesses of original building, optimizing functional layout, line, color, and material so as to create a perfect combination of aesthetics and practicality. His works harvested many domestic and overseas contest awards.

代表作品

成都中德英伦联邦A区5#楼3302户型、东莞龙泉豪苑9#2101样板间、贵阳乐湾国际A-3示范单位、江西南昌正荣御湾D1样板房

Representative Works

Chengdu British Ville A-5#3302 Show Flat, Dongguan Lungchuen Villas 9#2101 Show Flat, Guiyang Happy Valley International A-3 Show Flat, Nanchang Zhenro Bay Mansion D1 Show Flat

▼ 成都中德英伦联邦 A 区 5# 楼 3302 户型
Chengdu British Ville A-5#3302 Show Flat

▼ 东莞龙泉豪苑 9#2101 样板间
Dongguan Lungchuen Villas 9#2101 Show Flat

▼ 贵阳乐湾国际 A-3 示范单位
Guiyang Happy Valley International A-3 Show Flat

▼ 江西南昌正荣御湾 D1 样板房
Nanchang Zhenro Bay Mansion D1 Show Flat.

年度最佳样板房空间（大户型）
THE BEST SHOW FLAT SPACE 2014 (LARGE HOUSE TYPE)

北京万科北河沿甲77号样板房
VANKE BEIHEYAN PALACE 77 SHOW FLAT, BEIJING

颁奖词
Award Words

气势磅礴，空间区隔得当
Great momentum, with appropriate space partition

业主：北京万科企业有限公司　Client: Vanke Group Beijing Co., Ltd.
开发商：北京新铭房地产开发有限公司　Developer: Beijing Xin Ming Real Estate Development Co., Ltd.
项目地址：北京北河沿大街甲77号　Location: No. 77 North He Yan Avenue, Dongcheng District, Beijing
室内设计：上海朱周空间设计咨询有限公司　Interior Design: Vermilion Zhou Design Group
景观设计：上海陌尚景观设计有限公司　Landscape Design: Shanghai MS Design Co., Ltd.
建筑设计：清华大学建筑设计研究院　Architecture Design: Architectural and Design Institute of Tsinghua University Co., Ltd.
项目面积：400多平方米　Site Area: 400 m²
设计时间：2013年　Design Time: 2013
竣工时间：2014年1月　Completion Time: January 2014
采编：张培华　Contributing Coordinator: Zhang Peihua

康建国 董事、总经理
金盘地产传媒有限公司

这个空间采用了很好的框架理念，并且可以自由设计，给人以大气的感觉。

Kang Jianguo Director、General Manager
Kinpan Estate Media Ltd.

The space adopts a frame structure concept, which is beneficial for free design, leaving a feeling of grandeur.

嘉 宾 点 评
Honored Guest Comments

项目将传统文化与现代建筑艺术相结合,给人以大气磅礴感。大堂以廊柱、屏风、景墙布置,划分出多个子空间,营造生活私密感;可以自由分割的框架结构展现出含蓄低调的风格,承载了主人的理想和需求。

设计背景

项目概况

万科·北河沿甲77号位于北京市中心地段,毗邻故宫和圣地佛堂嵩祝寺,是北京中轴线旁的宝地。项目是坐落在北京人文中心区域的文化艺术大宅,设计旨在从项目地域周边历史和地块自身丰富的多元文化中挖掘题材,探索新中式景观的精神内涵,打造具有新时代意义的高端会所住宅。

设计理念

项目将传统与现代建筑艺术特点相结合,整体分为门、园、楼三重空间。设计师采用新中式园林的空间设计手法,在空间序列上通过一条主要轴线的贯穿,层层递进,一步一景,形成一府,一院,一园,一胡同的空间格局。在重要的景观节点上打造静、清、雅三个特性,凸显出浓郁的人文气息。

入口

入口磅礴大气,取传统古兽,雕琢玉玺五枚,依列席厅前水岸。从入口大门设计,到仿拟山水的对照,再到"双狮滚绣球"等传统石雕作品,展现出项目的皇家尊贵气质、佛教文化意境和低调奢华品质,使整个项目仿若一件珍贵"藏品"而非"居所"。

The project integrates traditional culture and modern architectural art, leaving a feeling of grandeur. Pillars, screens and landscape walls fix up in the lobby to divide it into several sub-spaces, creating a feeling of private life; the frame structure, able to split freely, expresses restrain and low-key flair, which bears the ideal and desire of the host.

Designing Backgrounds

Overview

Vanke Beiheyan Palace 77 situates in the center of Beijing, close to the Imperial Palace, the Holy Buddhist Temple and the Song Zhu Temple, so it is a treasured place beside Beijing central axis. The project is a cultural art mansion in Beijing cultural centers. The design is aimed at mining subjects from surrounding history and multi-culture of the land itself, exploring the spirit of Chinese landscape connotation and forging a high-end club residence of new era.

Design Concept

The project integrates traditional and modern architectural art, and the whole space is divided into gate, yard and mansion three categories. The designers employ new Chinese garden space design methods to set a main axis to go through the overall space sequence, level by level and one step a scene, forming a spatial layout of one mansion, one yard, one park with one alley. Tranquility, litheness and grace are three features incarnating on crucial landscape nodes, highlighting intense humanistic breath.

Entrance

The entrance is majestic and grand. Five carving imperial jade seals in Chinese ancient animals' models are in array in the hall at waterfront. People can feel imperial noble temperament, Buddhist cultural conception and low-profile luxury quality from the design of entrance gate, artificial landscape and traditional stone carvings of "two lions playing ball". The overall project is as a treasured collection but a residence.

样板间 A 户型
Apartment Layout A

大堂

北河沿甲77号的室内旅程首先从气势磅礴的大堂开始。大堂布置廊柱、屏风及景墙，整个大堂空间被划分为多个独立子空间；为了营造主人日常生活的私密感，设计通过布置屏风、景墙，甚至灯光，来区隔每个人独立的行走流线。

室内设计

室内设计有别于北京市场上的其他高档住宅产品，力图展现出含蓄低调的风格，承载了主人会友、论道、观天下的理想。空间整体采用框架结构，可自由设计分割，房间可270°观景，巨幅落地大窗可远瞰西山，直面景山，近观紫禁城，远眺CBD，多功能满足主人需求。

The Lobby

To visit the Beiheyan Palace 77, people will see the majestic and grand lobby at first. Pillars, screens and landscape walls fix up in the lobby to divide it into several sub-spaces; and the screens, landscapes and lighting are used as partitions to form independent pedestrian circulation, creating a feeling of private life.

Interior Design

The interior design is different from other high-grade residence in Beijing, while it expresses restrain and low-key flair, bearing the host's ideal of association, discourse and observation. The whole space adopts frame structure which can split freely. The rooms is available for viewing 270° scenery and the huge floor-to-ceiling window is allow for overlooking the Xishan Mountain, the Jingshan Mountain, the Forbidden City and Beijing CBD.

样板间 B 户型
Apartment Layout B

年度最佳样板房空间（大户型）
THE BEST SHOW FLAT SPACE 2014 (LARGE HOUSE TYPE)

宁波财富中心示范单位
FINANCIAL CENTER SHOW FLAT, NINGBO

颁奖词
Award Words

扇形空间形态、合理的灯光照明、流动的线条设计形成了极富艺术魅力的互动空间

Circular sector space, rational lighting and fluid lines comprise a charming interactive space

室内设计： 本则创意（柏舍励创专属机构） **Interior Design:** BASIC CONCEPT(PERCEPTRON GROUP)
项目地点： 宁波市江东区 **Location:** Jiangdong District, Ningbo
项目面积： 约500 平方米 **Site Area:** around 500 m²
竣工时间： 2014年1月 **Completion Time:** January 2014
主要材料： 海浪石灰、黑金砂石、夹纱玻璃、不锈钢、透光板等
Major Materials: Lime, Black Gold Sand, Clip Yarn Glass, Stainless Steel, DIAFOS
采编： 谭杰 **Contributing Coordinator:** Tan Jie

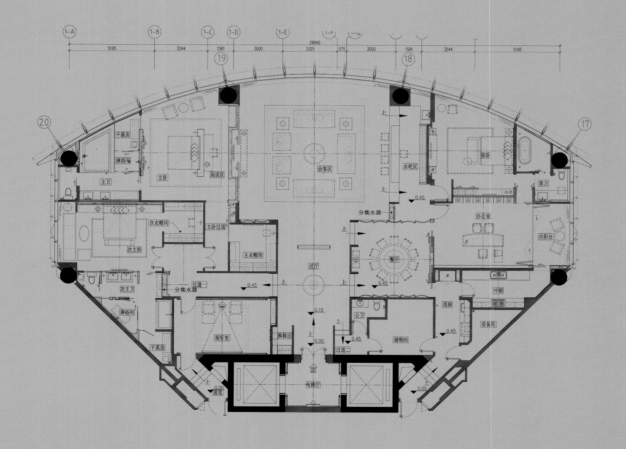

李仕泉 副总经理
金盘地产传媒有限公司

我觉得这个项目最大的特色是灯光照明运用得非常好，很有现代风格，而且设计师非常聪明，设计了很多流动的线条，形成了空间的互动。

Li Shiquan Director、General Manager
Kinpan Estate Media Ltd.

The best feature of the project is the lighting setting, offering a contemporary look. And designers smartly use plenty of fluid lines to outline an interactive space.

嘉宾点评
Honored Guest Comments

与建筑外形相吻合的扇形空间形态，配以最适合的照明，营造出一幅独特的照明意境图。空间整体色调沉稳大气，与众不同的天花板和墙身造型与精准的灯光照射形成强烈的对比，配以流动性的线条设计，让空间灵动、有趣，形成极富艺术魅力的互动空间。

设计背景

区位概况

宁波财富中心位于宁波市江东区，惊驾桥的北侧，滨江大道的东侧，江东北路的西侧。与宁波老外滩和美术馆隔江相望；南面重要历史建筑有庆安会馆和宁波书城。

建筑设计

本案的建筑外观设计非常富有视觉冲击力：在平面布局上，主楼与裙房形体自由，多个单体像中国传统的"玉石"撒在地块内，蕴含着"大珠小珠落玉盘"的诗情画意；在立面造型上，主楼形似含苞待放的郁金香，也像捧着明珠的手掌，看上去非常浪漫华丽。

钢结构

项目采用全钢结构，目前是宁波市高层建筑中施工难度较大，单体体量最大、高度最高的钢结构建筑，同时，幕墙采用单元式幕墙工艺也是宁波首次。

The circular sector space is identical with the architectural appearance, complementing splendid lighting, and the result is a particular lighting drawing. The general color is calm and grand. The modeling of ceiling and walls and precise light irradiation shape a sharp contrast, and coupling streaming lines, create an interactive space which is dynamic, interesting and fascinating.

Designing Backgrounds

Location

The Financial Center is in Jiangdong District, Ningbo. It locates at the north of Jingjia Bridge, east of Binjiang Avenue and west of Jiangdong North Road. It faces Ningbo Old Bund and Art Gallery across a river. Two significant historic buildings Keian Hall and Ningbo Book City are at its south.

Architecture Design

The project has an appearance full of visual impact: the main building and podiums are in free form on the plane layout, and several single buildings like jadestones scatter on the site, implying an ancient poetic illusion of pearls onlaying on jade plate; the main building facade is like a tulip bud or a pearl in hands, romantic and gorgeous.

Steel Structure

The project is all welded steel structure, and it is the most difficult construction high-rise building at present, also the biggest and highest steel structural architecture in Ningbo. Meanwhile, the unit curtain wall is firstly applied in Ningbo.

会客室——天花中央的弧形不锈钢造型与主幅墙的拉丝不锈钢相呼应，在灯光的照射下熠熠生辉
Living Room—The arc stainless steel modeling on the ceiling echoes brushed stainless steel on walls, and they are sparking under the irradiation of light

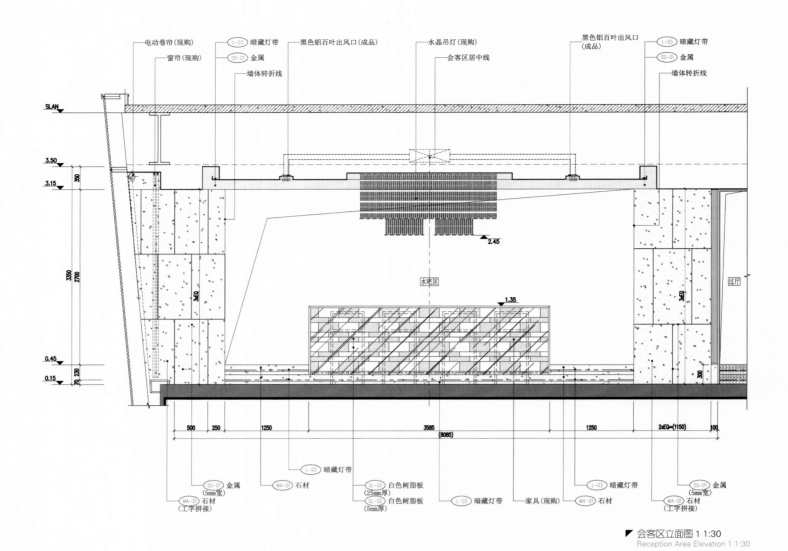

会客区立面图 1 1:30
Reception Area Elevation 1 1:30

会客区立面图 2 1:30
Reception Area Elevation 2 1:30

空间形态：扇形

为了与财富中心的建筑外形完美吻合，项目空间形态特设计成扇形，其中圆弧部分是最佳的观景位置。设计师根据项目的空间特色来划分功能区域，期望客户获得良好的生活体验；并且针对不同的空间装饰，选择最合适的照明以获得装饰上特有的情感升华，勾勒出一幅主次清晰、层次丰富、光线明亮的照明意境图。

设计理念

项目定位于现代风格，整体空间色调沉稳大气，彰显出主人不凡的气质。设计通过大面积发光，利用天花板和墙身的造型，精准的灯光照射，形成强烈的色彩对比，渲染了空间环境的尊贵，传递出一份厚重的高贵。

Space Form: Circular Sector

The circular sector space is identical with the architectural appearance, and the circular arc section is the best viewing position. Function zones are divided in accordance with spatial features to offer a good living experience to inhabitants; the light setting also accords with different spatial decoration to enhance emotional sublimation, outlining an ideal drawing enjoying clear primary and secondary, plenty gradation and bright light.

Design Concept

The project boasts its modern style, and the overall spatial hue is calm and grand, which manifests great taste of the host. It has a large area giving out light, and utilizes the configuration of ceiling and walls as well as precise light irradiation to shape a sharp contrast, rendering dignity and delivering a pronounced noble.

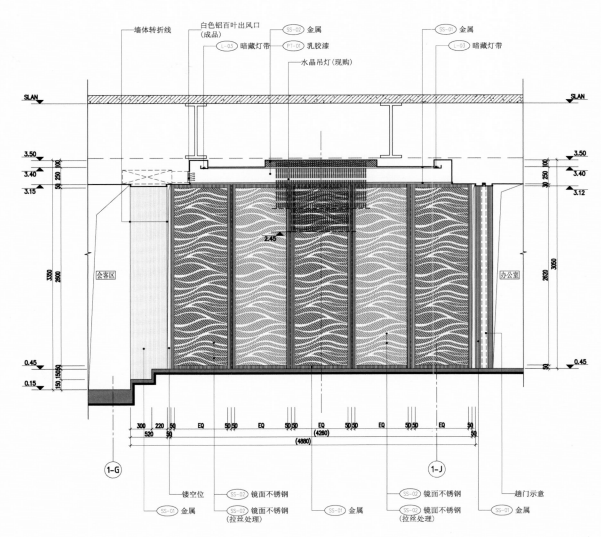

▶ 餐厅立面图 1 1:30
Dining Room Elevation 1 1:30

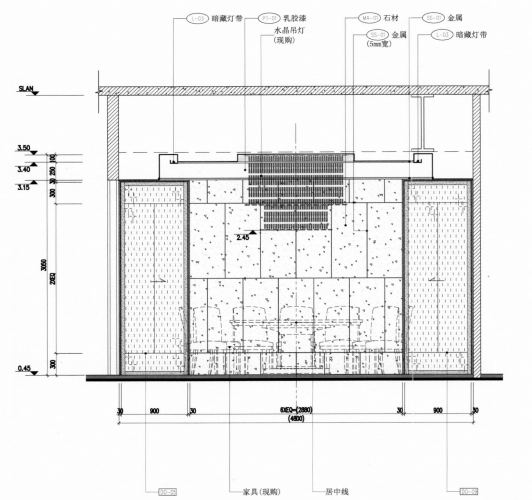

▶ 餐厅立面图 2 1:30
Dining Room Elevation 2 1:30

餐厅——层次分明的水晶吊灯与波纹大理石、动态的地毯相映成趣；
Dining Room—Leveling crystal pendant lamp, ripple marble and dynamic pattern carpet form a delightful contrast;

透光板——主要材料中的透光板特点：首先透光板表面不是玻璃或者石材，而是亚克力、聚碳酸酯等耐久性、透光率更强的材料；其次，透光板采用的是PP孔状芯材和特异结构芯材。透光板更轻盈，光学效果多样；力学结构合理，抗弯折能力强；有一定的隔音隔热特性；同时在采用不同表面材料后，可以显著增加抗UV和抗自然侵蚀的能力。

DIAFOS — the surface of the DIAFOS is not glass or stone, but acrylic and makrolon which have better durability and luminousness. It adopts PP cavernous core material and specific structural core material, so DIAFOS is lighter and has varied optical effects. DIAFOS boasts its reasonable mechanical structure, flexural folding ability and sound and thermal insulation characteristics. Different surface materials given, it would have outstanding ability of anti-UV and natural erosion resistance.

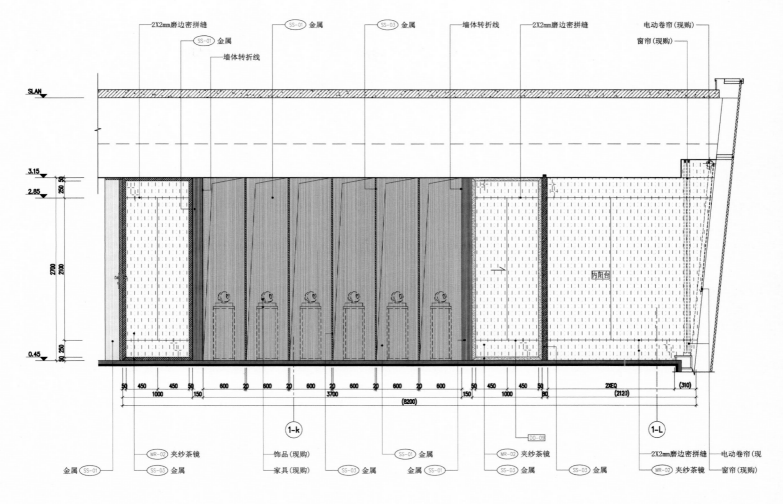

阅读室立面图 1 1:30
Study Elevation 1 1:30

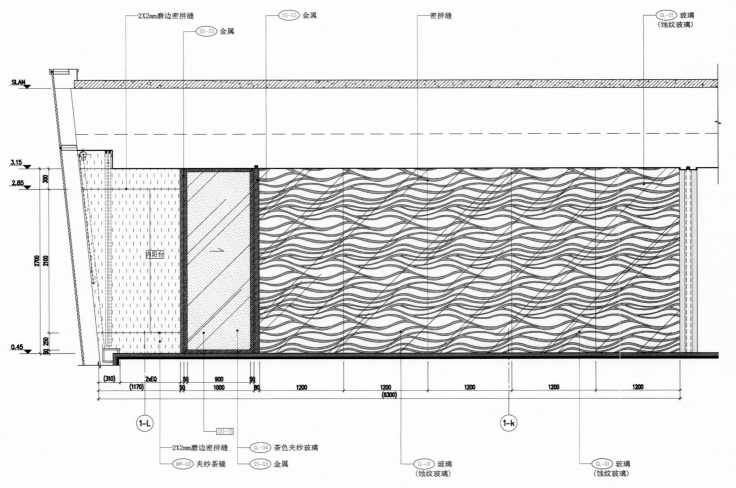

阅读室立面图 2 1:30
Study Elevation 2 1:30

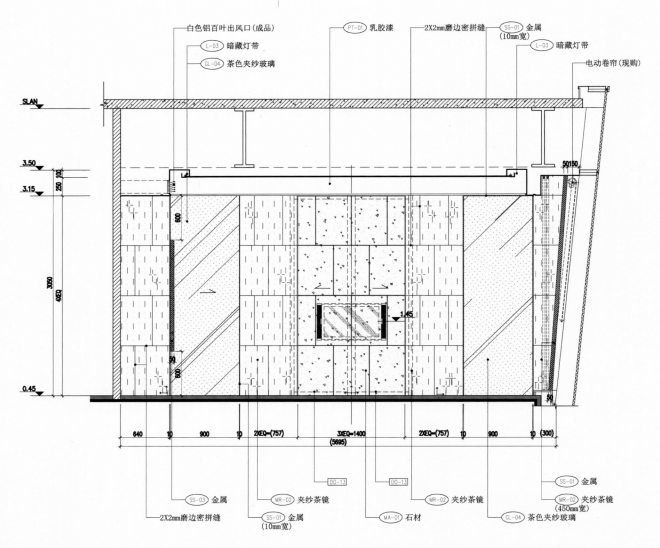

主卧立面图 1 1:30
Master Bedroom Elevation 1 1:30

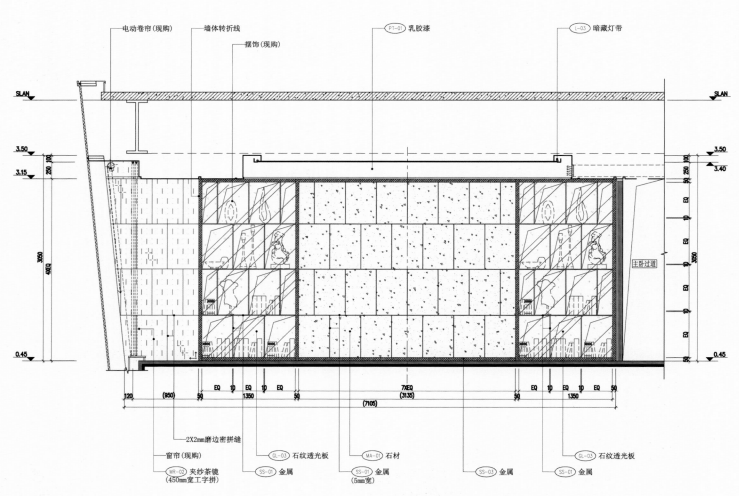

主卧立面图 2 1:30
Master Bedroom Elevation 2 1:30

梁智德 | Turkey Liang

梁智德,毕业于广东工业大学,有多年的室内设计及园林设计经验,比较擅长房地产类型及休闲娱乐类型项目。其坚持为人所用的原则,提倡功能与形式结合,合理利用资源,追求设计本质,发掘传统与现代的交汇点,让东方审美哲学的精髓片断留在人们记忆当中。

Turkey Liang, graduated from Guangdong University of Technology, has many years of interior and landscape design experiences, especially adept at real estate, leisure and recreation projects. He insists on design for people, and advocates integration of function and formation and rational utilization of resource, searches for the intersection of tradition and modernity and strives to insert the essence of oriental aesthetic philosophy into memory.

代表作品

保利德胜中汇花园8座303样板房、长白山池南区项目展示中心、佛山万科广场二期A区办公室、玖如堂C2示范单位

Representative Works

Poly Desheng Zhong Hui Garden 8-303 Show Flat, Changbai Mountain Southern District Exhibition Center, Foshan Vanke Plaza Phase II District A Office, City Center Park Townhouse C2 Show Flat

▼ 保利德胜中汇花园8座303样板房
Poly Desheng Zhong Hui Garden 8-303 Show Flat

▼ 长白山池南区项目展示中心
Changbai Mountain Southern District Exhibition Cente

▼ 佛山万科广场二期A区办公室
Foshan Vanke Plaza Phase II District A Office

▼ 玖如堂C2示范单位
City Center Park Townhouse C2 Show Flat

年度最佳样板房空间（中户型）
THE BEST SHOW FLAT SPACE 2014 (MEDIUM HOUSE TYPE)

佛山怡翠宏璟样板间
EMERALD COLLECTION SHOW FLAT, FOSHAN

颁奖词
Award Words

现代主义的建筑形体理念在空间内的完美应用
Modernism architectural form concept perfectly applied in space

开发商: 广东能兴房地产开发有限公司 Developer: Nenking (Holdings) Group Co., Ltd.
设计公司: 佛山硕瀚设计有限公司 Design Company: HON Design
主创设计: 杨铭斌 Chief Designer: Yang Mingbin
项目地址: 广东省佛山市禅城区 Location: Chancheng District, Foshan, Guangdong
项目面积: 93平方米 Site Area: 93 m²
设计时间: 2013年5月~6月 Design Time: May to June 2013
竣工时间: 2013年12月 Completion Time: December 2013
主要材质: 木饰面、铁方管、乳胶漆、镜面、木地板等
Major Materials: Wood Veneer, Iron Square Tube, Emulsion Varnish, Mirror Plane, Wood Floor
采编: 谢雪婷 Contributing Coordinator: Xie Xueting

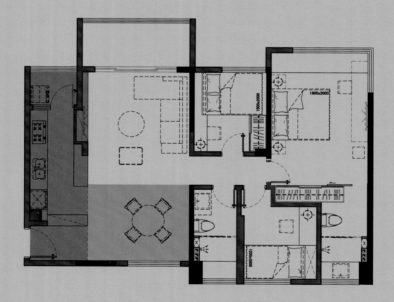

韩 松 总经理

深圳市昊泽空间设计有限公司

作为主力户型，佛山怡翠宏璟项目更具有对普通家庭的可推广性。

Han Song General Manager

Horizon Space

As anchor house type, the Emerald Collection show flat can be used for reference to ordinary families.

张 宁 总设计师

广州集美组室内设计工程有限公司

让我比较感动的是佛山怡翠宏璟项目。它整个设计看不出很豪华的感觉，却符合消费者的价值观，作为主力户型能够控制在这样的设计层面，让人们直接感受到这种户型的恰当性，也让我们在10平方米的空间可以清晰地看到设计师的核心想法。

Zhang Ning Chief Designer

Guangzhou Newsdays Interior Design & Construction Co., Ltd.

I am moved by Emerald Collection show flat project, because it is not luxury but meeting the value of intended customers. As anchor house type, people can feel the appropriateness obviously, which reflects the designers' main idea in the 10 m² space.

嘉 宾 点 评
Honored Guest Comments

现代主义建筑的形体穿插、特意的空间设置使整个空间流动起来，纵横交错的垂直水平立面建构出空间形体的简洁与纯粹。主人对现代主义的钟爱通过素白色的空间色调、灯饰与家具的搭配来彰显，低调而恬淡。

设计背景

区域概况

怡翠宏璟位于佛山核心地段——亚艺湖成熟豪宅片区，毗邻季华路与南海大道，畅享"四纵、九横、两环"公路网的便捷，轻松穿梭禅城、桂城、顺德，交通十分便捷。

规划设计

项目建筑面积约17万平方米，首期开发住宅约8万平方米。产品线丰富，涵盖24席亲水护城别墅、80~130平方米小高层舒适洋房及200平方米高层大平面楼王等多种创新形态。

设计理念

整个空间设计奉行现代主义建筑的形体理念，通过形体的穿插，设计师的特意设置使空间流动起来，显得通畅、动感。简单的立面与精致的材质，纵横交错的垂直与水平立面建构出形体的简洁与纯粹。空间里的形体构成，模糊各区使用功能空间的定义，营造宽广的空间视野；并巧妙地利用镜面反射，让空间得以延续，使室内空间得到很大的张力。

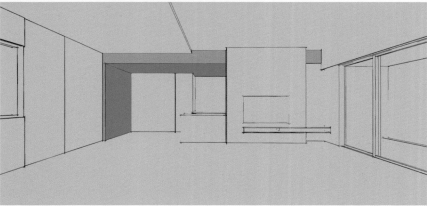

细节设计

有了空间的定义，设计师在选择灯饰与家具搭配的同时，展示出主人对现代主义的钟爱与品位；棉麻质感的沙发布艺，古色古香的木地板、木饰面静躺在素色纯白空间里，特别能衬托出灯饰与家具的质感，恬淡却不乏灵气。

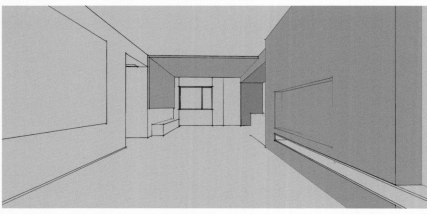

The crisscrossing modernism shape and special space arrangement make the space fluid; crossing vertical and horizontal facade structures create a pure and neat appearance. The pure white space, supplementing with lights and furniture, expresses host's love for modernism, simple but no lack of smart feelings.

Designing Backgrounds

Location

The project locates in the core area in Foshan, belonging to Yayi Lake district, a mature luxury house district and close to Jihua Road and Nanhai Avenue. It enjoys a convenient road system of "four south-north roads, nine east-west roads and two ring roads", so it is easy to go to Chancheng District, Guicheng District and Shunde District.

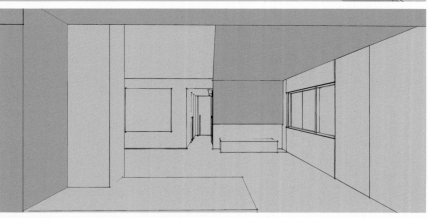

Planning Design

The building area of the whole project is 170,000 m^2, and the first developed residence covers 80,000 m^2. The project has plentiful formats, including 24 waterfront villas, small high-rise foreign-style houses of 80 m^2 to 130 m^2 and high-rise large flat buildings of 200 m^2.

Design Concept

The entire space design is inspired by the shape of Modernism architecture. The crisscrossing structure is designed expressively to make the space fluid. Basic facade, delicate texture, and crossing vertical and horizontal structure, all together create a pure and neat appearance. In the aspect of space function divisions, the definition is blurred to build a wide space vision. The use of mirror reflections further extends the interior space.

Design Details

Based on the clear space orientation, designers select lights and furniture to present client's love and taste for Modernism. Especially, the quality of lights and furniture is shown by the choice of pure and white colors, simple but no lack of smart feelings.

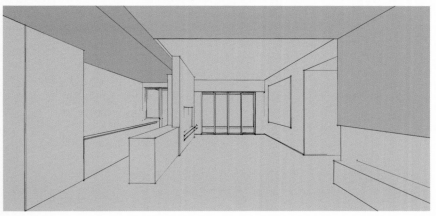

乳胶漆——水分散性涂料，它是以合成树脂乳液为基料，填料经过研磨分散后加入各种助剂精制而成的涂料。乳胶漆具备了与传统墙面涂料不同的众多优点，如易于涂刷、干燥迅速、漆膜耐水、耐擦洗性好等。

Emulsion Varnish — an aqueous dispersion coating, which takes synthetic resin emulsion as basic material, and after grinding then joins all kinds of additives to refine. Emulsion varnish has many advantages against other traditional wall coating, for example, it is easy to brush, quickly drying off, water resistant and washable.

杨铭斌 | Beni Yeung

杨铭斌,硕瀚创意设计研究室创始人及总设计师,以建筑空间的思维,强调空间内外的高度连接与流动性,追求空间本质的生命力,并率先提出将空间、人文、品位、氛围融合的概念。

Beni Yeung, the founder and chief designer of HON Design, he values spatial thinking and the connection and fluidity between internal and external spaces; he pursues intrinsic vitality of space; and he is the first designer proposes the concept that space, humanity, taste and ambiance should be integrated as one.

所获荣誉 Awards & Honor

第22届APIDA亚太区室内设计大赛样板房类别十大入围作品

第19届APIDA亚太区室内设计大赛住宅类别十大入围作品/购物空间类别铜奖

现代装饰国际传媒奖年度最具潜力设计师

CIID50位中国优秀青年设计师

第十七届中国室内设计大奖赛住宅类别银奖

22nd Asia Pacific Interior Design Awards Top 10 Nominee Works of Show Flat

19th Asia Pacific Interior Design Awards Top 10 Nominee Works of Residence / Bronze Award of Shopping Space

Modern Decoration International Media Award The Most Promising Designer

17th China Institute of Interior Design Silver Award of Residence

公司简介

"用心生活,用爱设计"是硕瀚设计的思维态度。硕瀚设计始终坚持为客户创造价值为宗旨,团队拥有国际化的视野及敏锐的市场触觉,针对不同行业、不同客户挖掘设计的核心价值,以新锐的理念,严谨的出品作为准则。

Design Company Profile

"Live with heart and design with love" is the thinking attitude of HON Design who insists on creating value for clients as objective. The design team has international vision and acute market sensitivity. They design according to different industries, different customers' core values, and use cutting-edge design ideas to produce prudent works.

代表作品

广东佛山碧水湾样板房、摩根国际写字楼会所、日本寿司连锁店等

Representative Works

Foshan Garden Riveria Show Flat, Morgan Centre Club, Japanese Sushi Multiple Shop

广东佛山碧水湾样板房
Foshan Garden Riveria Show Flat

年度最佳样板房空间（中户型）
THE BEST SHOW FLAT SPACE 2014 (MEDIUM HOUSE TYPE)

中山时代倾城四期04户型样板间
TIMES KING CITY PHASE IV 04 SHOW FLAT, ZHONGSHAN

颁奖词
Award Words

传统与现代、雅致与奢华并存的居住空间
A coexistence of tradition and modernity, elegance and luxury living space

项目地址： 广东省中山市沙溪镇 **Location:** Shaxi Town, Zhongshan, Guangdong
室内设计： 广州共生形态工程设计有限公司 **Interior Design:** C&C Design
主创设计： 彭征 **Chief Designer:** Peng Zheng
设计团队： 谢泽坤 **Design Team:** Xie Zekun
项目面积： 116平方米 **Site Area:** 116 m²
主要材料： 大理石、烤漆板、实木地板、硬包、墙纸、不锈钢、玻璃等
Major Materials: Marble, Lacquer Board, Solid Wood Floor, Hard Roll, Wallpaper, Stainless Steel, Glass
采编： 谭杰 **Contributing Coordinator:** Tan Jie

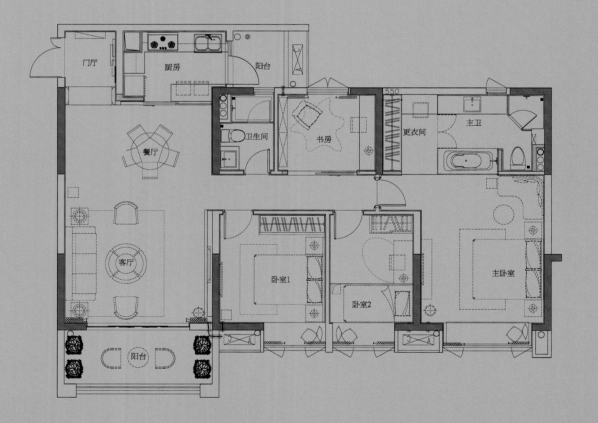

康建国 董事、总经理
金盘地产传媒有限公司

项目采用了蒙太奇般的设计手法，整体呈现出来是一个比较低调的居住空间。

Kang Jianguo Director、General Manager
Kinpan Estate Media Ltd.

The project takes Montage design skill to present a low-profile living space.

嘉 宾 点 评
Honored Guest Comments

项目在欧式新古典的框架下，融入大量的米色面料和镜面材质，展现出优雅、低调的现代居住空间。平面布局保留了原来的建筑格局，在塑造空间轴线的大原则下，适当地开放空间，保证功能布局合理和空间的流动性。

设计背景

区域概况

项目位于中山沙溪镇黄金通道博爱一路，紧邻岐江商圈，是中山市政府规划的居住中心之一，地处未来城市CLD的核心位置，地理位置非常优越。作为已开放了三期的项目，小区配套成熟、完善，景观绿化宜人，视野也较为开阔。

设计构思

针对此样板房户型是同期开放的户型中最大的，设计呈现了一个三代同堂的家庭组合。项目采用古典符号的装饰，犹如电影片段的闪回，现代与传统如同蒙太奇般交相辉映，营造出一个优雅、低调的居住空间。

A large amount of beige fabric and mirror materials are used in European neo-classic style to express an elegant and low-profile living space. The plane layout keeps the principle of shaping spatial axes, then to open space appropriately and guarantee rational function layout and walking circulation.

Designing Backgrounds

Location

The project is at Bo'ai Yi Road, the main road in Shaxi Town, Zhongshan. It is close to Qijiang business zone, and is one of the municipal planning residence centers. It occupies the core area of the future Central Living District. The property has finished three phases, so the community has mature and complete supporting facilities, pleasant landscape and broad vision.

Design Idea

This is the biggest one among the show flat house types, so the design presents a family house of three generations. The project takes classic signal ornaments, like movie clips flashback, making modernity and tradition mingle in montage way, and creating an elegant and low-profile living space.

设计手法

项目提炼和简化了欧式新古典的框架，融入米色面料和镜面材质，柔软细腻的面料给居家增添了几分温馨；镜面在空间中折射出绚丽的光影，为整个空间注入活力，展现出雍容与舒适、雅致与奢华并存的现代居住空间。

设计细节

厨房

整个空间的平面布局保留了原建筑四房两厅的格局。设计在塑造空间轴线的大原则下，拓宽了厨房的展示面，同时新增了早餐台，厨房的趟门因此解放出来，成为了"移动的风景"。

书房

设计师将其中一个采光面较小的卧室改为独立式书房，并整合书桌和书柜，将朝向过道的墙体打开，增加展示面的同时起到缩短和丰富走廊感官意向的作用。在走廊的尽头，设计将房门设置为端景，以此来强化轴线关系。

主人房

主人房被打造为一个酒店式的大套间，在保证了私密性的同时，最大程度地开放空间，既保证了视野的通透性，也保证了空间的流动感。独立更衣间和落地浴缸更增添了空间的尊贵感。推开房门，惬意的转角沙发背倚着浓浓英伦风味的大笨钟背景，茶几上的咖啡杯和书本仿佛一下又把人们带回了那惬意的伦敦时光。

烤漆板——木工材料的一种。它是以密度板为基材,表面经过六至九次打磨(不一样的厂家次数不同,次数越多工艺要求越高,成本也就越高)、上底漆、烘干、抛光高温烤制而成。主要用于橱柜、房门等。其优点有:颜色鲜艳;且选择多样;可配以多种搭配方式,颜色相间;容易清洁与打理;对增大厨房的空间有一定补光作用。

Lacquer Board — is a kind of wood material. It takes density board as basic material, and has following manufacturing processes: six to nine times burnish (the times are varied with different manufacturers; the more times the higher of its crafts and costs), painting undercoat, drying, polishing and baking in high temperature. Lacquer board is mostly used in cabinet and doors, because it has many advantages: bright colors, plenty choices, many collocation ways, color alternation, and it is easy to clean and maintain, and also can supplement light to enlarge kitchen space.

Design Method

The project refines and simplifies the European neo-classic frame, and adds beige fabric and mirror materials to increase the warm and sweet ambiance for the house. Radiant lighting reflecting from mirrors makes the space revived. A coexistence of tradition and modernity meanwhile elegance and luxury living space is witnessed.

Design Details

Kitchen

The whole plane retains the original four rooms two halls layout. The design keeps the principle of shaping spatial axes, then widens the kitchen's display panel and adds a breakfast counter. The adjustment makes the kitchen liberate from a fixed space, becoming a movable scene.

Study

Designers change the bedroom with small lighting surface into an independent study. The desk and bookcase are in recombination, the wall facing the aisle is burst through, which increases display panel and shortens and enriches sensory effect of the aisle. At the end of the aisle, a door is set as a side view to strengthen axis relations.

Master Bedroom

The master bedroom is as a hotel-type suite. It not only ensures privacy, but opens space to the extreme, so the room has transparent vision and fluent space. Independent cloakroom and bathtub increase noble feeling. Open the door, a corner sofa is against a British Big Ben background, with coffee cups and books on tea table. The setting brings people back to cozy London momentos.

彭 征 | Peng Zheng

彭征，共生形态合伙人，董事，设计总监。广州美术学院设计艺术学硕士，曾任教于中山大学传播与设计学院、华南理工大学设计学院。关注城市化进程中的当代设计，主张空间设计的跨界思维，设计作品具有较强的建筑感和现代简约的风格。

Peng Zheng, the Design Director and a Copartner of C&C Design, got Master degree of Fine Arts in Guangzhou Academy. Previously he was a guest lecturer of the School of Communication and Design in Sun Yat-Sen University as well as School of Design in South China University of Technology. He focuses on the contemporary design in urbanization process, advocates the cross-discipline thinking in space design, and his works feature strong sense of building and international style.

所获荣誉 Awards & Honor

美国室内设计年度最佳大奖、意大利 A 'DESIGN AWARD 银奖铜奖、香港亚太室内设计大奖、金堂奖、现代装饰国际传媒奖、艾特奖等国际和国内设计大奖

ID+D Annual Best Award, A' Design Award & Competition Silver Award and Bronze Award, Asia Pacific Interior Design Awards 2013 in Hongkong, JinTang Prize, Modern Decoration Awards, Idea-Tops Awards etc.

公司简介

"共生形态"是广州本土的一个新锐的精英设计团队，他们专注于地产项目的室内设计和整体软装服务。他们以"共生"为企业的核心理念，并为致力于当代中国设计面貌的成形而努力。

Design Company Profile

C&C is a cutting-edge design team in Guangzhou, and provides professional interior and overall soft decoration service for the real estate club. We are interested in all the connotation of the phrase "Cooperate & Coexist", and devote to the shaping process of the present China.

代表作品

南昆山十字水生态度假村、凯德置地御金沙、万科金色领域、珠江科技数码城、时代地产中心

Representative Works

The Crosswaters Ecolodge & Spa · Bamboo Villa, CapitaLand Dolce Vita, Vanke Golden Domain, Pearl River Technology Digital City, Times Property Center

▼ 南昆山十字水生态度假村
The Crosswaters Ecolodge & Spa · Bamboo Villa

▼ 2013 广州国际设计周共生形态馆
2013 Guangzhou International Design Week C&C Pavilion

▼ 南海万科金色领域售楼部室内设计
Wanke Golden Field Sales Center Interior Design

年度最佳样板房空间（中户型）
THE BEST SHOW FLAT SPACE 2014 (MEDIUM HOUSE TYPE)

济南建邦原香溪谷二期D5户型样板间
TOSCANA HOLIDAY PHASE II D5 SHOW FLAT, JINAN

颁奖词
Award Words

将样板间展示与接地气的生活化场景相结合，呈现别致现代生活
The display of the show flat combines with life scenes, presenting a unique contemporary life

开发商： 济南建邦置业有限公司　**Developer:** Jian Bang Real Estate
项目地址： 济南长清大学科技园大学城海棠路5555号
Location: No.5555 Haitang Road, University Town, Changqing University Technology Park, Jinan
设计公司： 成象空间设计　**Design Company:** Imaging Space Planning
设计师： 岳蒙　**Designer:** Yue Meng
设计时间： 2013年　**Design Time:** 2013
竣工时间： 2013年10月　**Completion Time:** October 2013
采编： 陈惠慧　**Contributing Coordinator:** Chen Huihui

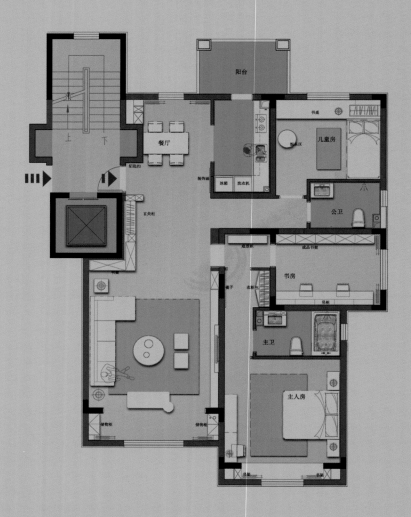

新浪地产网　**Dichan.Sina.com.cn**

设计重点回归到家的温暖与稳重，着力营造一种踏实安稳的归属感。
The key point is back on the warmth and steady of home, and the design strives to create a stable sense of belonging.

嘉宾点评
Honored Guest Comments

设计选择了忠实于原建筑空间的布局,在平民化的生活场景中挖掘生活魅力。简单而温暖是设计对于家的理解,深蓝与绿色的非常规组合打造出奇妙的空间效果,恰到好处的留白彰显设计功力,软装与硬装的搭配相得益彰,现代、清新、雅致,又不乏真实。

设计背景

该样板房所属的项目位于大学科技园区,周边高校林立,人文气息浓厚。项目定位为刚需家庭居住,户型面积为130平方米,简单方正。甲方对样板房的设计要求是尽量为客户提供较强的参考性与可复制性,要尽量接近人们真实的生活状态,呈现出人格化的效果。

设计思路

房子所承载的是家的幸福,设计希望塑造一个温暖的家庭氛围,在简单之中诠释幸福。幸福来自生活点点滴滴的小感动,来自每个角落中家人的欢笑记忆。设计提取空间中的阳光与爱,凝聚成充满生命活力的天然气息,令家人所经过的每一个角落都留下快乐的回忆。

空间设计:清爽雅致

设计整体以现代简约风格为主,颜色以白色与蓝灰色为基调,简洁大方且不失优雅。空间用色朴素而大胆,正对入户的是一幅抽象创意画,灰色的基调,跳动的画面仿佛走入一个寂静而又生动的世界。客厅活泼的果绿色,使整个空间瞬间充满盎然生命之感。主卧沉稳的咖啡色与灰色混油面板的搭配是对一天生活沉淀的真实写照。

The design reserves the original space layout, and excavates life charm from ordinary life scenes. Simple yet warm feeling is an interpretation of home; the unconventional combination of blue and green forms a wonderful space; precise blank reflects the capability of the designers; soft and hard decorations complement each other perfectly, making a modern, fresh, elegant yet no lack of real.

Designing Backgrounds

The project locates in Changqing University Technology Park, so it is surrounded by universities and full of humanistic breath. The project is built for rigid demand families. The show flat is of 130 m^2 area, simple and foursquare. The client proposes a requirement that the design should be a good reference and duplicable for intended customers, so it must be close to the real life status and present hommization.

Design Idea

A house bears the happiness of home, and the design wishes to produce a warm ambiance. While happiness is from little things in real life and the family sweet memories of every corner, so the design borrows the sunshine and love to form a vivid living space, making every corner have a happy memory of every family member.

Space Design: Fresh and Elegant

The design takes modern minimalism style, and chooses white and blue-gray as keynotes, concise and elegant. The space colors are plain yet bold. Open the door, an abstract painting in gray keynote jumps into sight, and the saltant picture brings people into a tranquil yet lively world. The living room decorated with yellowish green ornaments, fills the space with lively breath. The master bedroom uses stable coffee and gray mix oil panel, which is a mirror of everyday life deposit.

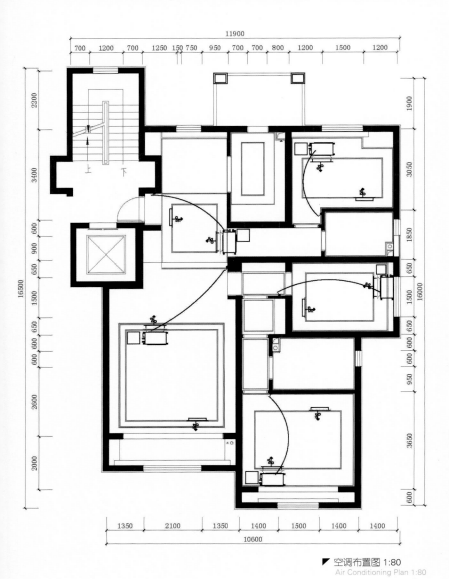

▼ 空调布置图 1:80
Air Conditioning Plan 1:80

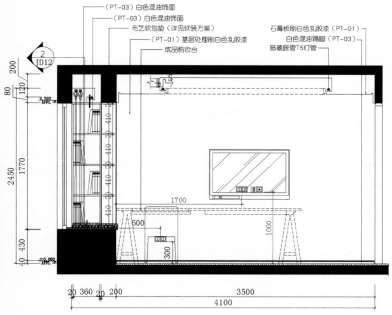

主卧室立面图 1 1:35
Master Bedroom Elevation 1 1:35

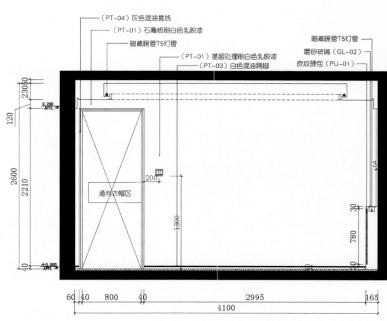

主卧室立面图 3 1:35
Master Bedroom Elevation 3 1:35

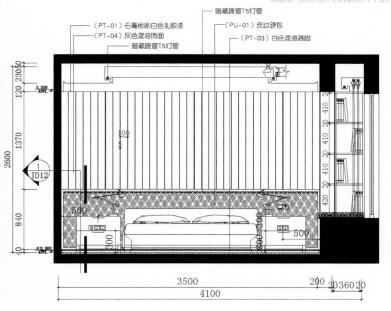

主卧室立面图 2 1:35
Master Bedroom Elevation 2 1:35

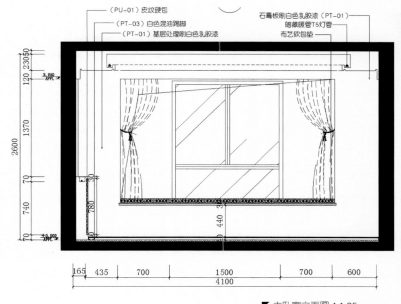

主卧室立面图 4 1:35
Master Bedroom Elevation 4 1:35

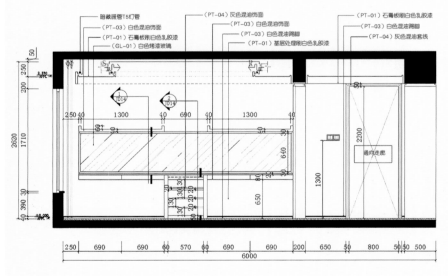

▼ 书房立面图 1 1:35
Study Elevation 1 1:35

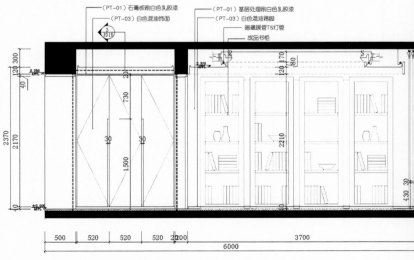

▼ 书房立面图 2 1:35
Study Elevation 2 1:35

岳蒙 | Yue Meng

岳蒙，亚太建筑师与室内设计师联盟理事，亚太酒店设计协会理事。现任深圳成象空间设计公司总经理，济南成象空间设计公司总经理，北京逸品成象空间设计公司设计总监。

Yue Meng, an APDF Director and Asia Pacific Hotel Association Director, who is the General Manager of Imaging Space Planning in Shenzhen Branch and Jinan Branch and the Design Director in Beijing Branch.

所获荣誉 Awards & Honor

2011年亚太设计精英邀请赛办公室类优胜奖

2011年金堂奖年度优秀别墅设计奖

2013年"时代楼盘第八届金盘奖"总决赛 最佳样板间设计

2013年金堂奖年度优秀样板间/售楼处设计

2011 Asia Pacific Design Awards for Elite Office Winning Prize

2011 JinTang Prize Good Design of Villa

2013 Times House 8th Kinpan Awards Final The Best Show Flat Award

2013 Jin Tang Prize The Best Show Flat Award, Jin Tang Prize The Best Sales Center Award

公司简介

成象空间设计是一家主要服务于地产项目的室内设计公司，向客户提供使其商业空间价值提升的服务。"成象空间设计的样板房最懂销售"是客户们的一致反馈。

Design Company Profile

Imaging Space Planning is an interior design company mainly in the service of real estate projects, and we offer customer service for improving the value of commercial space. We get unanimous feedbacks from clients "Imaging Space Planning's show flats are most helpful to sales".

代表作品

济宁绿地国际城A2户型样板间、临沂正直花园示范单位A2、临沂正直花园示范单位B1、山东泰安中齐国山墅165 m²中式样板间等

Representative Works

Green Land International Community A2 Show Flat, Linyi The Master Park A2 Show Flat, Linyi The Master Park B1 Show Flat, Zhong Qi Guo Shan Villa 165 m² Chinese-style Show Flat

▼ 济宁绿地国际城 A2 户型样板间
Green Land International Community A2 Show Flat

▼ 临沂正直花园示范单位 A2
Linyi The Master Park A2 Show Flat

▼ 临沂正直花园示范单位 B1
Linyi The Master Park B1 Show Flat

▼ 银丰唐郡 14# 别墅示范单位
Beautiful Place 14# Villa Show Flat

年度最佳样板房空间（小户型）
THE BEST SHOW FLAT SPACE 2014 (SMALL HOUSE TYPE)

佛山保利西雅图8栋A3样板房
POLY SEATTLE 8-A3 SHOW FLAT, FOSHAN

颁奖词
Award Words

直线、直角与三原色的空间组合，向抽象艺术家蒙德里安致敬
A space of straight lines, right angles and three-primary colors, it pays our respects to abstract artist Mondrian

开发商: 保利华南实业有限公司 **Developer:** Poly South China Holdings Co., Ltd.
项目地址: 广东省佛山市南海区 **Location:** Nanhai District, Foshan, Guangdong
设计公司: 广州道胜装饰设计有限公司 **Design Company:** DS Design
主创设计: 何永明 **Chief Designer:** He Yongming
项目面积: 58平方米 **Site Area:** 58 m²
主要材料: 银河世纪大理石、爵士白大理石、瓷砖、复合地板、黑色镜面不锈钢、防火板等
Major Materials: Galaxy Marble, Volakas Marble, Ceramic Tile, Composite Floor, Black Mirror Steel, Fire-proof Plate
设计时间: 2013年5月 **Design Time:** May 2013
竣工时间: 2013年9月 **Completion Time:** September 2013
摄影师: 彭宇宪 **Photographer:** Peng Yuxian
采编: 谢雪婷 **Contributing Coordinator:** Xie Xueting

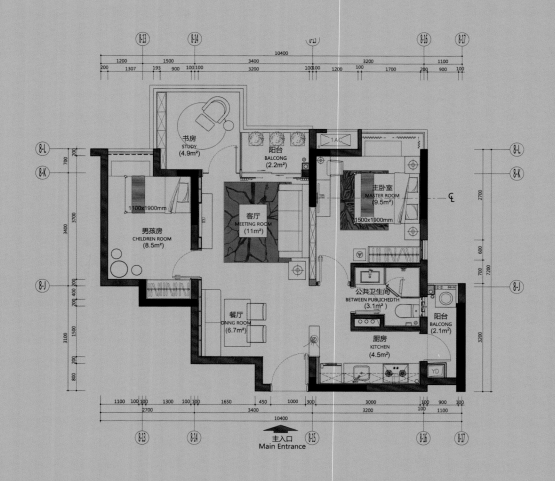

搜狐焦点网
本方案运用"蒙德里安红黄蓝的直线美"为元素，采用大小不等的红、黄、蓝色块创造出强烈的色彩对比，同时塑造出稳定的平衡感。

Focus.cn
The project makes use of the elements of Mondrian red, yellow, blue and straight lines, and sets various color lumps to create intense color contrast, in the mean time, forms a stable balance feeling.

嘉宾点评
Honored Guest Comments

设计用直线、直角以及红、黄、蓝三原色，打造出极富韵律感的空间氛围，用最基本的几何元素创作出冲突与和谐并行不悖的空间意象，在以黑、白、灰为大背景的空间中致敬蒙德里安，以抽象艺术元素为简约的生活空间增添艺术气息。

设计背景

区位交通
项目位于广佛几何中心，地处三大商圈黄金交汇点，尽享通达四方、立体多维的便捷交通，5分钟接驳广州南站，与老城区、珠江新城形成"三站共生"城市格局。

楼盘概况
项目拥有自身商业配套约3万平方米，拥有1000平方米泳池，超过1300平方米的会所，设施完善，是超大体育休闲社区，并有全国首创二维码园林。首批单位将主打以刚需自住为主的城市精英群体。

建筑风格
保利西雅图以现代的建筑风格，融合新古典的经典建筑格调，打造出具有大气之风的经典之作。

The design uses straight lines, right angles and three-primary colors to form a rhythmic space. Basic geometric elements create a space in a balance of conflict and harmony. Under the black, white and gray background, abstract art elements add an artistic ambiance to the space, which pays our respects to artist Mondrian.

Designing Backgrounds

Location
The project sits in the center of Foshan, and it is the intersection of three business districts. It enjoys convenient transport conditions, only 5 minutes to Guangzhou South Station. It unites Guangzhou old town and Zhujiang New Town, forming an intergrowth layout.

Overview
The project has 30,000 m² business supporting, 1,000 m² swimming pools, over 1,300 m² clubs. It is a super sports leisure community, and has the first two-dimension code garden in China. The first residences are for rigid demand customers and city elites.

Architecture Style
Poly Seattle adopts modernism style and integrates neo-classic style to present a grand and classic works.

彼埃·蒙德里安（1872—1944）——荷兰画家，风格派运动幕后艺术家和非具象绘画的创始者之一。他认为艺术应脱离自然的外在形式，以表现抽象精神为目的，即"纯粹抽象"。他崇尚直线美，主张透过直角可以静观万物内部的安宁，所以他的很多作品以几何图形为基本元素，挖掘色彩与线条中的秩序与韵律感，充满了音乐性。

Piet·Mondrian (1872-1944) — a Netherland painter. He is an artist of de Stijl (the Neo-Plasticism) and also one of the founders of Abstract Painting School. He believed that art should break away natural external form but to perform an abstract sprit as purpose, which was "purely abstract". He admired the beauty of straight lines, and proposed to use right angles to watch the tranquility of everything, therefore many of his works took geometrics as basic elements to explore the order and rhythmic feeling in colors and lines, full of music beauty.

复合地板——又叫强化地板，一般是由四层材料复合组成，即耐磨层、装饰层、高密度基材层、平衡(防潮)层。合格的强化地板是以一层或多层专用浸渍热固氨基树脂覆盖在高密度板等基材表面，背面加平衡防潮层、正面加装饰层和耐磨层经热压而成。优点有：耐污、抗酸碱性好，免维护；富有弹性、防滑性能好；耐磨、抗菌，不会虫蛀、霉变；尺寸稳定性好、不会受温度、潮湿影响变形等。

Composite Floor — also called laminate floor, usually is compounded of four layers—a wearing layer, a deco layer, a high density base material layer and a balance moisture-proof layer. Qualified composite floor is hot pressed by a layer or multi-layers high density base material surface infused solid amino resin, a balance moisture-proof layer on reverse side, a deco layer on observe side and a wearing layer. It has following merits: resistance to staining, good resistance to acid alkali, maintenance-free, high resilience, non-slip performance, wearing resisting, antibiosis, no worm-eaten and mildew, good stability, non-effect of temperature and moisture deformation.

设计主题：蒙德里安的音乐画面

空间灵感来自于荷兰的抽象艺术家蒙德里安的作品。画面由长短不同的水平线和垂直线构成大小不一的正方形和长方形，并以粗黑的交叉线将他们分开，在正方形周围用各种长方形穿插，那些原色以及黑、灰、白的对比、排列就像音符旋律中的变化。

空间细节

家具上运用原木来增加空间的自然与和谐感，传达出对悠闲自由的生活方式的追求。饰品、挂画、地毯等都细腻地延续着蒙德里安元素，色彩大胆跳跃。儿童房中，有趣的墙贴和地上的玩具呼应了儿童开朗活泼的性格。主卧在色彩丰富的墙面和地毯上，用素雅的浅灰色来中和过度，让空间丰富的同时保持稳重不浮躁，使整个空间的色彩在冲突中和谐。

Design Theme: Mondrian Music Picture

The space design inspiration is from the works of Netherlander Mondrian, an abstract artist. There are various quadrates and rectangles composed by horizontal and vertical lines, and with bold black lines to divide them. The rectangles are surrounding the quadrates. The three-primary colors and black, gray and white colors form intense contrast, just like the variation of melody.

Space Details

Raw wood, used in furniture, expresses nature and harmony, and reflects the host's desire for free and leisure life. The Mondrian elements are applied in ornaments, pictures and carpets, with bold and dynamic colors. In children's room, interesting wall posters and toys on ground reveal the kids' lively character. In the master bedroom, elegant and quiet light-gray neutralizes multi-colored walls and carpets, making the room in a balance of richness and prudence and the space color in a balance of conflict and harmony.

何永明 | Tony Ho

何永明,设计总监,以现代主义精神与热情为设计注入完美无瑕的风格和创新能量。透过整合建筑、室内设计、视觉图像和室内布置,每一次新作皆创造出独特的感官魅力与欢愉的空间气氛。

Tony Ho, the Design Director in DS Design. He is good at using modernism spirit and passion to create designs of refined style and innovative power. Through combining architecture, interior design, visual pictures and indoor layout, his every new works is of unique charm and full of joyful ambiance.

所获荣誉 Awards & Honor

2009年荣获APIDA第十七届香港亚太室内设计大奖荣誉奖
2012年荣获金外滩最佳饰品搭配优秀奖
2012年荣获金指环-IC@WARD2011全球室内设计办公空间设计大奖
2013年荣获艾特奖-最佳商业空间设计入围奖及最佳样板房设计入围奖
2013年荣获CIID-优秀青年室内设计师及杰出青年室内设计师
2014亚太室内设计精英邀请赛入围奖
2014第十二届现代装饰国际传媒奖 年度样板空间大奖
2014中国室内设计年度评选金堂奖 年度设计公益奖

2009 APIDA 17th Hong Kong Asia Pacific Interior Design Awards Honor Award
2012 The Golden Bund Awards the Best Decorations Collocation Excellence Award
2012 The RING IC@WARD International Design Awards Interior Office Space Design Award
2013 Idea-Tops the Best Commercial Space Design Nominee Award & the Best Show Flat Design Nominee Award
2013 CIID Outstanding Young Interior Designer & Prominent Young Interior Designer Titles
2014 Asia Pacific Interior Design Awards For Elite Nominee Award
2014 The 12th Modern Decoration International Media Award Annual Sample House Space Award
2014 JinTang Prize China Interior Design Awards Public Welfare Award

公司简介

广州道胜设计有限公司及何永明设计师事务所一直追求创意、实用、优质的设计。公司的设计理念为"以人为本,美化空间",令三维的设计产生思维的体验,达到"人居合一"的完美境界。

Design Company Profile

Guangzhou DS Design Company and Tony Ho Architects Associates have been pursuing design of creativity, practical and high quality. They hold the concept "people oriented and beautifying the space", committing to make three-dimensional design produce intellectual experience and creating a prefect state of "habitation syncretism".

代表作品

心动城市—保利三山西雅图销售中心、利地产—繁花似锦,保利东湖林语大堂,南昆山会议中心,南亚小站餐厅,金叶香辣轩餐厅,乐嘉快餐厅,马拉爸爸餐厅,乐嘉茶艺馆,寿师傅日本餐厅等

Representative Works

Poly Seattle Sales Center, Poly Real Estate-Central Pivot Plaza, Poly East Lake Hall, Nan Kun Shan Conference Center, Banana Leaf Asian Café, Grand Restaurant & Pub, Rock Restaurant, Malay Papa Asian Café, Rock Tea House, Sushifu Sushi Teppanyaki

▶ 保利三山西雅图销售中心
Poly Seattle Sales Center

品质

一个好品质酒店不仅仅考虑的是好的材质与好的施工工艺,还要基于合理的规划与设计,以及酒店特有的属性等。在有形的设备设施达到一定标准后,考虑到消费者的需求与利益,为其提供一些无形服务更能决定一个好品质酒店的接待能力和档次。

艺术

充满艺术气息的酒店总是让消费者流连忘返,越来越多的酒店空间通过科学与艺术的结合获得灵魂和生命,有自己的主题,引领着时尚和潮流。

人居

从人居方面来讲,好的酒店空间首先要有充满人性化的设计,有好的空间布局,合理的客房布置,健全的功能配套,能够满足客户的居住要求。其次,要注重细节,对客房的控制、对尺寸的拿捏以及相应的配置,都体现出设计的细心与用心。

价值

酒店是人们旅游度假、体验生活、歇宿的重要载体。从价值的角度来讲,酒店要以人为本,在为消费者提供舒适、周全、方便的同时,也要让人享受到轻松愉悦、宾至如归的感觉。

Quality

A high-quality hotel should give consideration to three aspects: good materials and construction crafts, appropriate planning and design, as well as unique attributes of its own. When tangible facilities are ready, customers' demands and benefits should be taken account seriously, because certain intangible service is the key-point to decide the accommodation capacity and its level.

Art

An artistic hotel always makes customers enjoy themselves so much as to forget to leave. More and more hotels obtain their soul and life through combination of science and art. These hotels have their own themes and lead the fashion trends.

Habitat

From the aspect of habitat, a preeminent hotel space is full of humanized design in good spatial layout, rational guest room arrangement and wholesome functional supporting so as to satisfy residing demands. In addition, the controlling on details, guest room, scale and allocation reflects the designers' circumspection and intention.

Value

Hotel is an important carrier catering for tourist resort, life experience and making overnights stop. Hence hotels should be people oriented, providing comfortable, comprehensive and convenient services, in the meantime bring relaxing and pleasing feelings to make guests feel at home.

年度最佳酒店空间
The Best Hotel Space 2014

本类别包含了度假酒店、精品酒店和商务酒店。随着酒店行业竞争的不断加剧，越来越多的酒店空间设计者增强了对酒店发展空间环境，以及客户需求趋势变化的深入研究，不断寻求突破与创新，在注重设计的纯粹和文化定位的同时，也在追求和商业市场的结合度。

This chapter includes three hotel types of resort, boutique and business. Along with the increasing competition of hotel industry, more hotel space designers strengthen the research on environment of hotel development space and customers' demand trends, and in order to seek breakthrough and innovation, the designers highlight pure design and cultural positioning while search for the combination with business market.

厦门乐雅无垠酒店
Hotel WIND, Xiamen

成都钓鱼台精品酒店
The Diaoyutai Boutique, Chengdu

深圳四季酒店
Four Seasons Hotel, Shenzhen

曲阜香格里拉大酒店
Shangri-La Hotel, Qufu

年度最佳酒店空间（度假酒店）
THE BEST HOTEL SPACE 2014 (RESORT HOTEL)

厦门乐雅无垠酒店
HOTEL WIND, XIAMEN

颁奖词
Award Words

一个现代村落式的度假酒店，可沟通的居住氛围
A village-pattern resort hotel, a communicative residing atmosphere

开发商: 厦门乐雅酒店投资有限公司　**Developer:** LEYOINN Investment Co., Ltd.
项目地址: 厦门市思明区环岛干道黄厝段云岳山庄　**Location:** Yun Yue Heights, Island Road, Siming District, Xiamen
室内设计: 百伦设计　**Interior Design:** Balance Design
建筑设计: 上海间筑设计　**Architecture Design:** TEAM_BLDG
占地面积: 2 580 000平方米　**Site Area:** 2,580,000 m²
建筑面积: 21 270平方米　**Building Area:** 21,270 m²
主要材料: 天然柚木　**Major Material:** Natural Teak
采编: 谭杰　**Contributing Coordinator:** Tan Jie

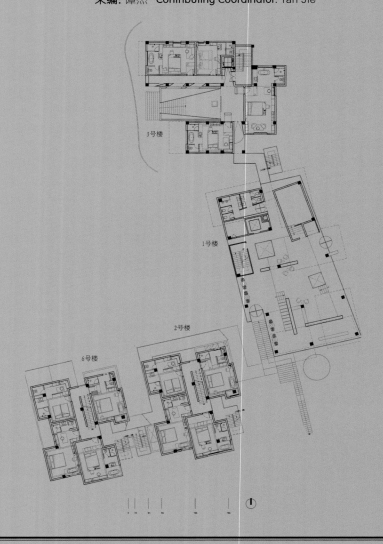

吴 滨 创始人
世尊设计集团

这个项目我比较喜欢，因为它对整个光影应用恰到好处，把度假酒店中客人的需求全表达了出来。

Wu Bin Founder
W+S Deco Group

I like this project a lot, and especially its application of light and shadow completely expressing the needs of guests in resort hotel.

嘉宾点评
Honored Guest Comments

无垠酒店为了营造独特的舒适感，最大限度地将房间所面向的美景通过方窗一一呈现，并且每个房间的风景都独一无二。设计通过不同材料的使用以表达清晰明确的功能分区，当代艺术空间丰富了入住者的体验与互动。

设计背景

区域特征

无垠酒店依山面海，位于美丽的海滨城市厦门的环岛干道上万石涌翠的云顶山畔，拥有43间优质客房及套房，客房面积由40至100平方米不等，满足不同居住需求。

规划设计

项目集合水疗、酒廊、健身瑜伽、创新美食、私家影院等功能于一体，是一个居住及当代艺术体验有机结合的高端城市度假酒店社区。

建筑设计

建筑上，无垠酒店借鉴闽南传统乡村的环境认知及以宗祠作为功能中心的聚落样式，由不同"几何"体块围绕着承担当代艺术展览的WE SPACE，堆叠组织而成一个现代立体村落。在与建筑外部保持良好私密性的同时，空间内部通过雨井、树井、风井等多重建筑器官的设置，促成内与外、人与环境、公共与私密的对话可能。

设计灵感

设计之初，设计公司希望项目能像一种流通的介质，不仅在山海间建立连接，也在人与人、人与环境间建立沟通，hotel WIND中的"WIND"便取意于此。而对于自然和环境的探索，也是"无垠"一词所希望传达的旅人精神。

The hotel WIND boasts its unique comfortable experience. Beautiful scenes can be seen as far as possible from rectangular windows, and every room has its own perspective. Various materials smartly divide functional zones, and artistic space enriches the sensation and interaction for guests.

Designing Backgrounds

Location

The hotel WIND, facing sea and against mountain, sits on Yun Ding Hill side, in coastal city Xiamen. The hotel has 43 excellent rooms and suites ranging from 40m² to 100 m², catering to different requests.

Planning Design

The hotel includes hydrotherapy, lounge, fitness yoga, gourmet food, private cinema, etc. It forms a high-end urban resort hotel community, organically combining residing and contemporary art experience.

Architecture Design

The hotel WIND borrows the environmental cognition of traditional Southern Fujian village and settlement pattern of ancestral temple as function center. Several geometric blocks embrace WE SPACE, the administrative lounge, piling up to a modern stereoscopic village. The building exterior keeps a good privacy while the interior space sets up rain well, tree well and air shaft as architectural organs to promote the communication between inner and outer spaces, human and environment, public and privacy.

Design Inspiration

The design company wants to build a hotel like a circulating medium, building communication not only between sea and mountain, but between people and people and environment and people. The word "WIND" is just like a medium to help people explore nature and environment, meanwhile expresses traveler spirits.

当代艺术体验

无垠酒店的当代艺术空间——WE SPACE，除了承担行政酒廊功能，还将通过定期的展览和沙龙，为入住者提供绘画、雕塑、多媒体艺术、观念艺术等多种艺术形式，成为拥有独立态度的艺术交流平台，当代艺术的体验更将深入酒店每一个空间。

功能分区

房间内，通过不同材料的使用表达出清晰明确的功能分区：

休息区——以柚木为主打材质，强调温暖舒适；

室内墙面——将房间所面向的美景通过方窗来一一展现，每个房间望出去的风景都各不相同，给客人留下独一无二的"明信片"；

客房——均有依据优质景观设计的开窗，阳台与露台更延伸了房间视野；

淋浴区——更多地与户外联通，室外圆形浴缸更可获得美妙的沐浴体验；

盥洗区——选用更回归本源的材质，凸显朴素洁净；

家具及灯具——皆由曾获德国红点设计奖的日本设计团队Balance Design为无垠量身设计。

Modern Art Experience

The hotel WIND modern art space—WE SPACE is an administrative lounge, while regular exhibition, salon will be held here. It is an independent art communication stage for drawing, sculpture, multi-media and conceptual art, etc., which makes the modern art ambiance penetrate in every corner of the hotel.

Functional Division

Different materials utilization divides function zones:

Resting area—teakwood as major material, highlighting warm and comfort

Interior wall—many windows like postcards decorated on interior wall, making guest rooms different from one another

Guest room—window, balcony and terrace are elaborately designed to enjoy broader view

Shower area—more chance to connect with outdoor space; round bathtub brings wonderful bathing experience

Bathroom area—selecting origin material to highlight homeliness and cleanness

Furniture and luminaires—customized by Japanese team Balance Design who won Red Dot Design Award

柚木——本案主要材料中的柚木是制造高档家具、地板、室内外装饰的好材料。柚木对多种化学物质有较强的耐腐蚀性，故宜作化学工业用的木制品。特别是运用于地板，耐腐、耐磨，光泽亮丽如新，花纹美观，色调高雅耐看，稳定性好，变形性小。

Teakwood — a great material used in top grade furniture, floor and indoor and outdoor decoration. It has strong corrosion resistance to a variety of chemicals, so it is suitable for chemical industrial wood products. It is more especially used on the floor, due to the ability of anti-corrosion, wear-resistance and gloss as bright as new. Teakwood has beautiful stripes and tinge, and is good in stability and deformability.

年度最佳酒店空间（精品酒店）
THE BEST HOTEL SPACE 2014 (BOUTIQUE HOTEL)

成都钓鱼台精品酒店
THE DIAOYUTAI BOUTIQUE, CHENGDU

颁奖词
Award Words

宽窄巷子里的精致下榻，最具中国元素的别样风情
An exquisite hotel in Kuan Alley and Zhai Alley, special amorous feelings of Chinese elements

业主：成都娇盈锦联投资有限公司　Owning Company: Chengdu Jiaoyin Jinlian Investment Co, Ltd.
酒店管理集团：钓鱼台美高梅酒店管理有限公司　Hotel Management Group: Diaoyutai MGM Hospitality
项目地址：四川省成都市青羊区宽巷子38~39号
Location: No.38-39 Kuan Alley, Qingyang District, Chengdu, Sichuan
设计公司：4BI建筑事务所　Design Company: 4BI Architecture Firm
设计师：Bruno Moinard　Designer: Bruno Moinard
建筑面积：16 000平方米　Building Area: 16,000 m²
采编：陈惠慧　Contributing Coordinator: Chen Huihui

新浪网

从建筑由里及表的层层相连，到室内室外空间的呼应转换，设计师以他独到的领悟，将生活展现得栩栩如生，洋溢着对真实生活的幻想，带给我们无限遐想。

Sina.com

The architecture is layer-upon-layer connected from inside to outside, and the indoor and outdoor spaces are mutual echoing and converting. The designer uses his unique insight to express a lifelike architecture, permeating a vision of real life and bringing infinite reverie.

嘉宾点评
Honored Guest Comments

项目处处彰显出中式风格和法兰西风情、地域特色与人文情怀的和谐交融，给客人带来具有时光穿越感的入住体验，更于细微之处透露着设计师对当地文化的独特理解。

设计背景

区域特征

酒店坐落于具有300多年历史的都市风情街区——成都宽窄巷子。宽窄巷子是成都的历史文化地标，它是成都超过2 300年建城历史的一个重要片断，至今仍保持着老成都市民的生活场景。在两座16 000平方米的中式庭院中，45间客房、4间餐厅酒廊及专属俱乐部分布得错落有致。

项目概况

酒店由主持设计了卡地亚品牌多家零售店面的法国顶尖设计师Bruno Moinard担纲设计。其主体由"宽庭""窄苑"两重院落构成，设有45间极具特色的客房，中国特色的庭院融汇了当今国宾馆的雍容气度，酒店装饰也融入了丰富多样的中国元素。

建筑设计

建筑设计保留了老建筑的中式神韵，古朴的灰砖凝聚了当地的历史气息。屋顶的优雅弧线，屋檐与门窗的中式细节，都体现出了地域特色与独特的人文情怀。

The hotel manifests Chinese style and France flavor in everywhere. Regional characteristics and humanities are integrated harmoniously, and guests would have an experience of traversing the time. The details disclose the designer's unique understanding into local culture.

Designing Backgrounds

Location

Chengdu Kuan Alley and Zhai Alley, where the hotel is situated in, are style streets with a history of 300 years, and also historical and cultural landmarks of Chengdu. They are a significant part of Chengdu's over 2,300 years long history. The lifestyle of old Chengdu is passed down in here. In two 16,000 m² Chinese courtyards, 45 guest rooms, four restaurants, lounge and VIP Club are well-proportioned.

Overview

Bruno Moinard, a top French designer who has designed many store fronts for Cartier, is invited to design the hotel. The hotel highlights two courtyards—The Mansion and The Courtyard, and includes 45 featured guest rooms, as well as Chinese-style courtyards, sharing the luxuriousness of The Diaoyutai State Guesthouse. The space is decorated with various Chinese elements, which makes guests feel welcomed upon their arrival.

Architecture Design

The architecture design retains Chinese style of old buildings, with pristine gray bricks drawing in the local historic breath. The graceful arc of roof and Chinese elements on eaves, doors and windows all reveal distinctive local features and profound humanistic feelings.

大堂酒廊的天幕设计
Lobby lounge

设计风格

中式的建筑风格与具有法兰西浪漫风情的家具相融合,彰显低调奢华的空间魅力。酒店还将法兰西的浪漫风情注入典雅幽静的东方庭院,将宽窄巷子300多年的人文气息与钓鱼台品牌高贵底蕴相融合,打造出精致优雅,又不失摩登风范的空间。

客房

酒店的45间客房或时尚大方、别具一格,或细致入微、精湛唯美。每间客房的屋顶都沿着原有屋顶的形状被设计成波浪形,增加了房间的高度,让客人的视野更开阔。每个房间都保留原有的中式窗户,增加了酒店的古典韵味。客人可以开窗远眺,欣赏老街区的建筑和美景;也可以推开房门,体验时光穿越的感觉。

特色功能区

御苑国宴餐厅——将传统的中国元素进行了国际化的表达,从木质拱形门通过即可到达宽敞的餐厅空间。

吊顶——沿着原有建筑物屋顶的形状用天然的木材装饰,搭配古朴简单的吊灯,细节之处尽显高端大气。

钓鱼台俱乐部——集美食、会议、娱乐为一体,展示成都最高端、最尊贵的聚会场所。

芳菲秀大堂酒廊——一廊三吧(雪茄吧、香槟吧和寿司吧)的功能组合。

东方餐厅有连通着窄巷子的户外阳台
KZ Restaurant & Grill

Design Style

French romantic flair is infused into the tranquil Oriental courtyard, combining the time-honored cultural touch of Kuan Alley and Zhai Alley with the noble connotation of Diaoyutai. The result is an exquisite and graceful, yet modern space.

Guest Room

The hotel has 45 guest rooms, fashionable and unique or elaborate and exquisite. Room roofs all along the original shape are designed in a wavy pattern, increasing the height of the rooms and making a broader view for guests. Chinese style windows are retained, strengthening classic feelings. Guests can open windows to enjoy the buildings and scenery of the old streets, or open doors to experience the feeling of traversing the time.

Featured Function Zone

Royal Court Restaurant—Traditional Chinese elements are expressed by an international way, and stepping through an arch wooden door a spacious dining space is witnessed.

Ceiling—Natural wooden adornments are along with the shape of original ceiling. Guests can feel a grand space from the plain and primitive pendant lamps.

Club—Gourmet food, conference and entertainment are integrated, it is the most high-end and noble meeting place.

Fangfei Garden—Cigar Bar, Champagne Bar and Sushi Bar are concentrated in this lobby lounge.

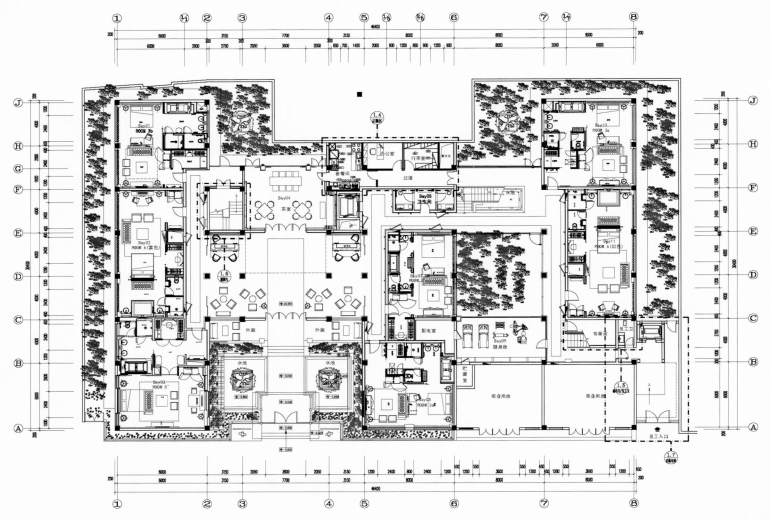

窄苑一层放大平面索引图 1:150
The Courtyard 1F Plan Index

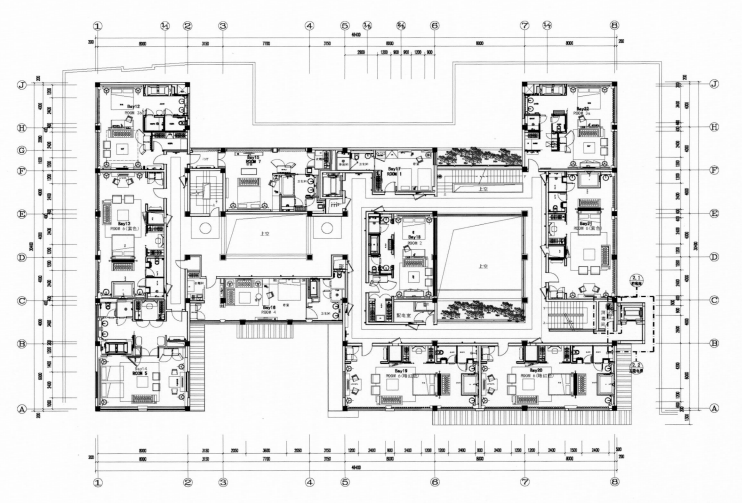

窄苑二层放大平面索引图 1:150
The Courtyard 2F Plan Index

年度最佳酒店空间（商务型酒店）
THE BEST HOTEL SPACE 2014 (BUSINESS HOTEL)

深圳四季酒店
FOUR SEASONS HOTEL, SHENZHEN

颁奖词
Award Words

现代艺术与传统元素完美融合，缔造典雅艺术空间
A perfect integration of modern art and traditional elements, creating an elegant artistic space

业主：卓越集团　Client: Excellence Group
项目地址：广东省深圳市福田区　Location: Futian District, Shenzhen, Guangdong
设计公司：HBA　Design Company: HBA
项目面积：80 000平方米　Site Area: 80,000 m²
开业时间：2013年　Opening Time: 2013
主要材料：玛瑙石、玻璃、大理石等　Major Materials: Carnelian Cobble, Glass, Marble
采编：谢雪婷　Contributing Coordinator: Xie Xueting

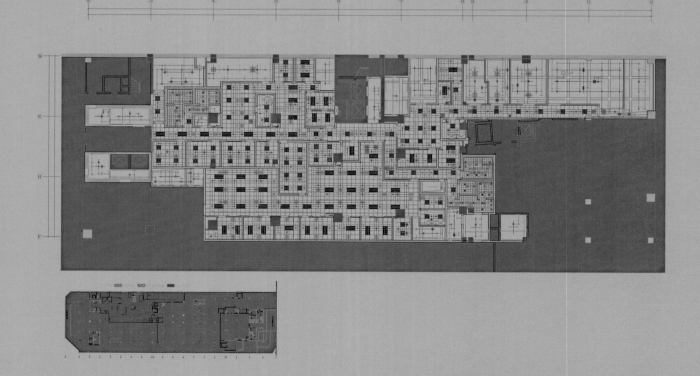

张 宁 总设计师
广州集美组室内设计工程有限公司

我选广东深圳四季酒店，因为它做得符合四季的主题，没有跑调。

Zhang Ning　Chief Designer

Guangzhou Newsdays

The reason why I chose Four Seasons Hotel Shenzhen is that the design pefectly conforms the theme of Four Seasons.

嘉 宾 点 评
Honored Guest Comments

在空间设计上，设计师巧妙融合传统与现代精髓，令酒店呈现出精品画廊般的典雅艺术气质。充满现代气息的材质结合带有现代抽象艺术感的纹理与图案，与贯穿酒店大堂、客房、餐厅等处的祥云等传统纹饰，碰撞出丰富而和谐的视觉效果。空间整体色调搭配适宜，给人放松愉悦的空间体验。

设计背景

酒店位于深圳福田区商业中心，是卓越时代广场二期综合开发项目的一部分，四周矗立着大型高档办公楼和三座购物商城。酒店的几何形建筑外观时尚耀目，打破了传统的直线型结构，以积木拼搭的形态出现。这样的建筑结构除了提供给客房完美的自然采光之外，更创造出一个露天平台，可一览城市风光，而且其建筑设计的现代艺术张力，也给空间设计奠定了基调。

功能概况

酒店楼高32层，拥有266间标准客房和豪华套房，并配备了四间高格调餐厅及完备的Spa服务等。酒店还设有不同样式、大小的宴会厅，主要分布于酒店3层、7层和29层。其中7层和29层的宴会厅以茉莉、玉兰、牡丹、莲花等花卉为主题命名，尽显中式美感。独一无二的别致户外平台，可举办酒会或婚礼庆典。

设计灵感

设计师灵感源自当地文化，巧妙融合传统与现代精髓，令酒店呈现出精品画廊般的典雅艺术气质，在深圳这个现代大都市中，为宾客创造出层次丰富的艺术体验与发现美的空间。

Designers smartly integrate traditional and modern essence to design the hotel space, making the hotel just like an art gallery full of elegant flavor. Modern materials, abstract texture and patterns as well as the auspicious cloud ornament adorned in lobby, guest room and dining hall stimulate a rich and harmonious visual effect. The color collocation is reasonable, bringing a relaxing experience.

Designing Backgrounds

The hotel locates at the commercial center of Futian District, and it is a part of Excellence Times Square Phase II. Large-scale high-grade office buildings and three shopping malls are around it. The geometric appearance is fashionable and bright, which gives up traditional pure line structure but adopts toy bricks pattern. This kind of architectural structure can offer not only a perfect natural lighting condition but also an open terrace to enjoy the scenery of the city. The building design expresses contemporary artistic sense, which sets the tone of the space design.

Function Overview

The hotel building has 32 layers, with 266 standard rooms and luxury suites, and equipped with four high quality restaurants and complete Spa services, etc. The hotel offers banquet halls in various styles and sizes, mainly distributed on the 3rd floor, 7th floor and 29th floor. Jasmine, orchid, peony and lotus are featured themes customized for banquet halls on the 7th and 29th floors, revealing the Chinese aesthetic flavor. The unique outdoor platform is convenient for cocktail party or wedding ceremony.

Design Inspiration

The design inspiration is from the local culture, and it brings a rich leveling artistic experience and a space to explore beauty in Shenzhen, a modern metropolitan city.

艺术空间

现代气息

迈过玛瑙石镶嵌的酒店门廊即是挑高的大堂，白色大理石锻造而成的几何形迎宾台与其背后的艺术墙呼应，带来静中有动的奇妙感受；大堂中央的水景带来视觉和听觉的双重享受；大堂另一头是一件布满整个墙面的艺术装饰，吸引宾客慢慢走入艺术长廊，长廊中是一系列精心挑选的瑰丽艺术作品。简约的线条，镜面材质，抽象艺术装饰，都让空间充满了现代艺术气息。

传统元素

大堂地毯中的祥云图案在客房的墙面中反复出现，令吉祥如意的美好寓意绵延不止。餐厅门口摆放着彩色骨瓷幻化而成的现代艺术作品，与垂挂于天花板的木质元素相互映衬。屏风和壁画也经常出现传统绘画中的写意风格，祥云、花鸟等都极富艺术美感，营造出别出心裁的设计感。

空间细节：宜人色彩搭配

迎宾大堂以墨色和暗金色为主色调，一踏入便有安逸奢华的美感。前台处以红色沙发点缀，增加了活力。客房以奶油色和米色为主色调，点缀以少许紫红色、灰绿色与柔金色，营造出轻松愉悦的氛围。深色木质床头柜和电视机柜搭配优雅的白色大理石台面，彰显超凡品味。米色搭配灰褐色的大理石内饰高贵典雅，与室外的自然元素相辅相成。

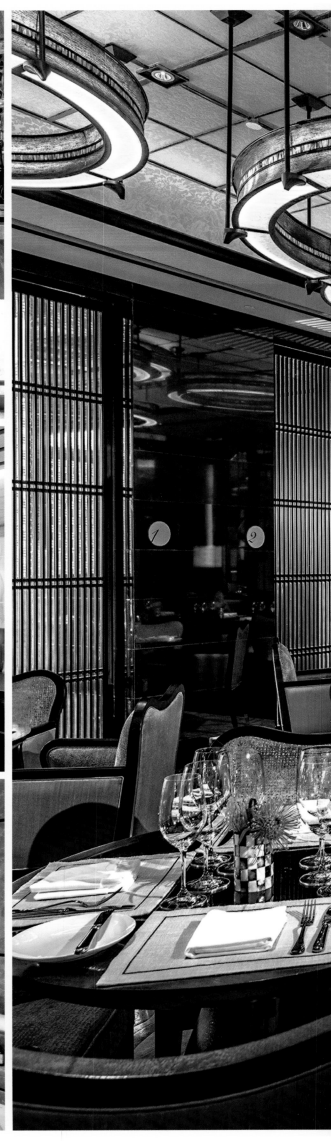

Art Space

Modern Ambiance

Stepping through the carnelian cobble decorated gate is the empty parlor height lobby. The geometric white marble reception desk echoes the lobby art wall; the waterscape in the middle creates visual and auditory enjoyment; in the other end of the lobby sets an art adornment spreading the whole metope, absorbing guests into the art gallery which is dressed selective artworks. The concise lines, mirror materials and abstract art ornaments all bring modern ambiance to the hotel.

Traditional Element

The auspicious cloud pattern is applied on lobby carpet and frequently presented in guest rooms, which lets the auspicious moral stretch long and endless. Colored bone-china artworks are placed at the door of the dining hall, echoing the wooden elements hanging on the ceiling. The screens and murals are in traditional Chinese freehand brushwork style, and the auspicious clouds, flowers and birds in painting are of extremely aesthetic feeling.

Space Details: Color Collocation

The reception lobby is in ink and dark blond hue, emitting out a relaxed and luxury feeling as guests step in. The red sofa beside reception desk adds dynamic atmosphere. While the dominant hue of guest room is cream and beige, with a little purplish red, gray green and gentle gold interspersed, creating easy and pleasing ambiance. On the white marble table board is dark wooden bedside table and TV bench, showing extraordinary taste. The beige and gray brown marble interior decoration is elegant, echoing the outdoor natural scenery.

年度最佳酒店空间（商务型酒店）
THE BEST HOTEL SPACE 2014 (BUSINESS HOTEL)

曲阜香格里拉大酒店
SHANGRI-LA HOTEL, QUFU

颁奖词
Award Words

新中式古典空间格调，承载浓郁的儒家情怀
New Chinese classic space, bearing profound Confucian feelings

业主：香格里拉酒店集团　Client: Shangri-La Hotel Group
项目地址：山东省曲阜市春秋中路　Location: Chunqiu Middle Road, Qufu, Shandong
室内设计：AB Concept / CCD　Interior Design: AB Concept / CCD
建筑面积：地上53 445平方米、地下21 387平方米　Building Area: over-ground 53,445 m², under-ground 21,387 m²
主要材料：铜、瓷砖、石雕、大理石、玻璃、木材等　Major Materials: Bronze, Ceramic Tile, Stone Carving, Marble, Glass, Wood
采编：吴孟馨　Contributing Coordinator: Wu Mengxin

搜房网

受儒家哲学的启发，设计团队独具匠心地将含蓄图案、丰富质感及对称美学融合为一，巧妙传递出儒家哲学的个中精髓。

Soufun.com

Inspired by Confucian philosophy, the design team smartly integrated implicit pattern, rich texture and symmetric aesthetics into one so as to deliver the essence of Confucian philosophy.

嘉宾点评
Honored Guest Comments

设计着力于呼应孔子故里的文化氛围，将空间主题定位于儒家文化，对六艺的诠释与儒家哲学的提炼赋予空间深厚的文化底蕴。除此之外，中国传统艺术也让空间更加具有表现力，青花、水墨、花鸟，中式意境在现代简约手法的表现下蔓延开来。

设计背景

区位文化背景

酒店坐落于孔子故乡、入选全国首批历史文化名城的曲阜，毗邻著名的孔庙、孔府与孔林。这样的区位优势让酒店拥有明确而清晰的文化定位——那就是打造以"三孔"文化为核心的高端文化商务酒店。

建筑外观

建筑外观用简约现代的方式传承了中式传统建筑风貌，两栋主体建筑宛如中式亭苑，附属建筑及庭院景观都借鉴了孔宅的形制，构成了一组独特的建筑群，既有现代商务酒店的高尚气派，又兼立足于孔子故里的文化底蕴。

The design echoes Confucius home village cultural ambiance. The space theme is positioned on Confucian culture, the interpretation of six classical arts and refining of Confucian philosophy, which makes the hotel space rich of cultural deposits. Beyond that, the Chinese traditional art enhances the expressive force of the hotel space. China flower, ink painting and abstract flowers and birds are presented by modern minimalist, expressing Chinese artistic conception.

Designing Backgrounds

Regional Culture Backgrounds

The hotel locates in Confucius home village Qufu, one of the historic cities first selected in China, and it is close to Confucian Temple, Confucius Family Mansion and Confucian Residence. The location advantage is beneficial for forming clear cultural position—focusing on "three Confucian" culture high-grade business hotel.

Architecture Appearance

The architecture appearance adopts modern minimalist to inherit Chinese traditional architectural style. Two main buildings are like Chinese courtyard, the affiliated architecture and landscapes borrow the shape and structure of Confucius mansion, forming a unique building group. The hotel boasts its modern business noble style and Confucius home village cultural deposits.

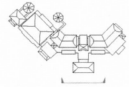

▼ 东南立面图 1
Southeast Elevation 1

▼ 东南立面图 2
Southeast Elevation 2

功能区域

酒店拥有三百余间豪华客房和套房,均可俯瞰河景、城市风光与生机勃勃的中式花园。酒店设有多处就餐区,可品尝纯正的孔府佳肴;健身俱乐部拥有先进完备的运动设施,设有室内恒温泳池、蒸气浴、按摩浴、桑拿、水疗中心和网球场;酒店的会议及宴会区是该区域面积最大、功能最全的举办会议的场所。

设计理念:儒家六艺

酒店的内部设计采用了儒家"礼、乐、射、御、书、数"的六艺理念,从走廊到客房,从逸水吧到孔咖啡厅,从健身俱乐部到演讲厅,处处都体现出儒家文化的情怀。设计结合现代手法,将中国传统儒家文化进行提炼与表达,营造出文化气息浓厚,又不失豪华舒适的空间。

Function Zone

The Shangi-La Hotel has 300 luxury guest rooms and suites, and every room can overlook the river, the city scenery and vibrant Chinese garden. It has several dining areas where guests can taste authentic Confucius family cuisine; fitness club has advanced and complete athletic facilities: indoor heated pool, vapor bath, massage bath, sauna, spa and tennis court; the conference rooms and banquet halls in the hotel are the largest and most complete function meeting venue in the district.

Design Concept: Confucian Six Classical Arts

The interior design adopts Confucian six classical arts "rite, music, shoot, ride, literacy and math" concept. Guests can get the Confucian culture feelings in everywhere, such as corridor, guest room, water bar, Café, fitness club and lecture hall. The design unites modern techniques to refine and express Chinese traditional Confucian culture, creating a luxury and comfortable space of rich cultural breath.

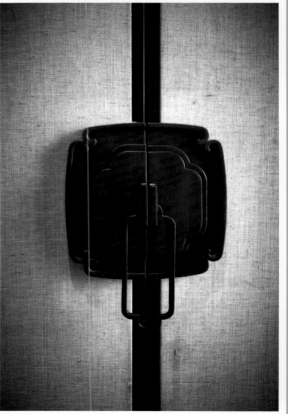

空间细节

接待大堂：儒家哲学

"天圆地方"与"中庸之道"是儒家哲学的重要内容，接待大堂对这两点进行了充分地演绎：圆形吊灯配方形晶串，方形大门饰圆形镂雕，殿柱左右对称，深红色漆质屏风将空间分割组合，前台的背景是情意盎然的巨幅多媒体百梅图。

餐饮区：古典情调

餐厅的内部装饰色彩以柔和的金色、深木色和橙色相呼应，古典与现代气息并行不悖。墙面由丝绢装饰，上有手工刺绣的中国花鸟庭园图案，寓意吉祥美满。5个包房和开放就餐区由玻璃墙面隔开，上绘水墨山水花鸟图。除此之外，酒店随处可见以简约写意形态出现的中国传统文化符号，如青花、木雕等装饰，营造出宁静简约的古典情调。

Space Detail

Reception Lobby: Confucian Philosophy

"Round heaven and square earth" and "golden mean" are important contents in Confucian philosophy, and the reception lobby makes a full presentation of these two contents in decoration: round pendant lamp adds square crystal beads, square gate adorns round hallow engravings, symmetric columns stands at sides, dark red painting screen partitions the space, and the background of the reception desk dresses huge multi-media drawing of hundred wintersweets.

Dining Area: Classical Sentiment

The interior decoration in dining hall is in gentle gold, dark wood and orange hue. Classical and modern elements run parallel. The wall is adorned with tiffany which embroidered with Chinese abstract flowers and birds, yards and pavilions, expressing auspicious luck. Five private dining rooms and open dining area are divided by glass walls which are painted with ink landscape painting. Beyond that, Chinese traditional cultural signals, such as Chinese flowers and wood carvings are witnessed everywhere, creating a tranquil and minimal classical sentiment.

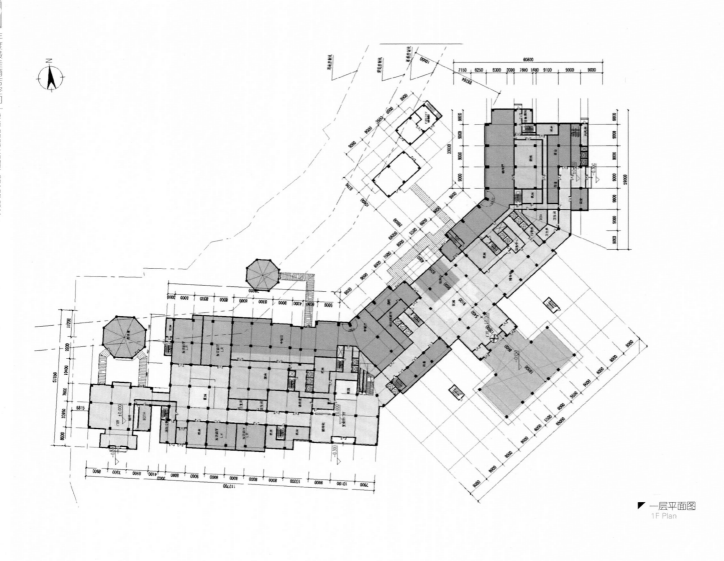

▼ 一层平面图
1F Plan

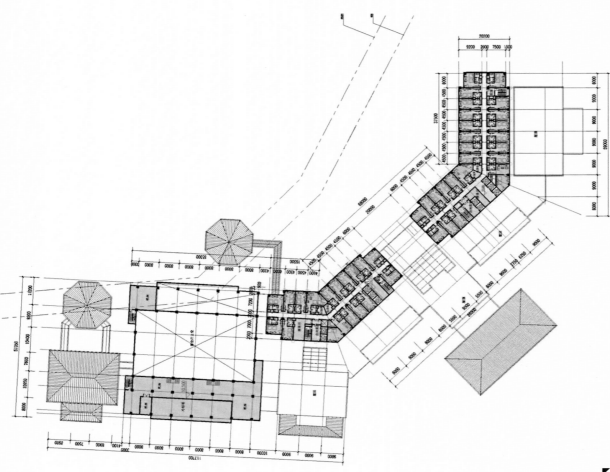

▼ 三层平面图
3F Plan

品质

好品质的售楼会所,不仅要功能齐全,而且能够因人而异,满足不同人的多种需求。在视觉效果上,设计特点突出,外表结构明显,内部装饰独特。不仅有充满艺术感的线条、夺目的色彩,还具有不息的生机和丰富想象,能让人产生购买冲动。

人居

功能分区成熟的售楼会所,拥有成熟的文化定位、消费群切分与完整的商业模式,能让目标受众明确自己的实际需求。无论是教育资源、生活方式,或是居住的优越性、外部商业的优越性,都能通过好的功能布局和配置,使购买者有个良好的体验氛围。

艺术

从艺术的角度来讲,售楼会所要有强烈的视觉和示范作用,用艺术的方式营造"代入感",能够让顾客在放松、舒适的环境氛围中,接受潜移默化的楼盘熏陶。

价值

优秀的售楼会所,无论是从完整性上、文化上,还是从商业模式上、语言上、功能上,都要经得起推敲。另外,从产品角度来讲,是一个优秀的设计;从消费者角度来讲,售楼会所的示范性设计要能够刺激消费欲望。

Quality

A high-quality sales center should have complete functions and satisfy various demands simultaneously. From the aspect of visual effect, it owns unique characteristics in apparent appearance and inner decoration. Artistic lines and eye-dazzling colors are expressed in vitality form and abundant imagination, inducing buyers' purchasing desire.

Habitat

A sales center with mature function division usually enjoys clear cultural positioning, classification of consumers and self-contained commercial mode so as to make receivers understand their own requirements. Alright functional arrangements in education resources, life-style and superiority of habitation and external business can bring a favorable experience to buyers.

Art

From the perspective of art, a sales center should have visual impact and demonstration effect to bring consumers a relaxing and comfortable atmosphere and be unconsciously edified by the property.

Value

A preeminent sales center can stand up to scrutiny in completeness, culture, commercial mode, design language and function. As a product, firstly it is an excellent design; secondly its demonstration effect should stimulate consumers' purchasing desire.

售楼会所能够让客户对项目的品位以及档次产生最直观的感受，不仅能够代言楼盘，更能充当一种信息沟通使者的角色。因此，优秀的售楼会所，无论是设计装修、视觉效果，还是售楼部的功能定位、商业手段，一定要做到独创新意，拥有市场中其他项目不具备的设计，效果具有不可复制性和仿效性。最重要的是还应该让业主体会到整个小区的文化氛围、消费群定位等。

The taste and level of a project can aware intuitively from a sales center which not only stands for the property but is an information messenger. Therefore, a preeminent sales center should be meticulous in its decoration, visual effect, function positioning and business methods; in the same time it should be more novel than other projects in the market in non-replication and non-imitation; and the most important of all, it should lead the buyers to receive the cultural ambiance of the community and their positioning.

年度最佳售楼会所空间
The Best Sales Center Space 2014

黄山悦榕庄度假别墅区展示中心
Banyan Tree Resort Villa Show Center, Huangshan

海南清澜半岛示范区
PE ninsula Demonstration Zone, Hainan

广州萝岗绿地中央广场售楼处
Guangzhou Greenland Central Plaza Sales Center

年度最佳售楼会所空间
THE BEST SALES CENTER SPACE 2014

黄山悦榕庄度假别墅区展示中心
BANYAN TREE RESORT VILLA SHOW CENTER, HUANGSHAN

颁奖词
Award Words

清、雅、朴、华，古代徽派美学融入现代生活
Feature in clear, refined, plain and unadorned, ancient Hui-style aesthetics blend in contemporary life

开发商：悦榕集团　Developer: Banyan Tree Group
项目地点：黄山市黟县　Location: Yi County, Huangshan
设计公司：HMD建筑设计有限公司　Interior Design: HMD
占地面积：3 000平方米　Site Area: 3,000 m²
建筑面积：600平方米　Building Area: 600 m²
采编：康欣　Contributing Coordinator: Kang Xin

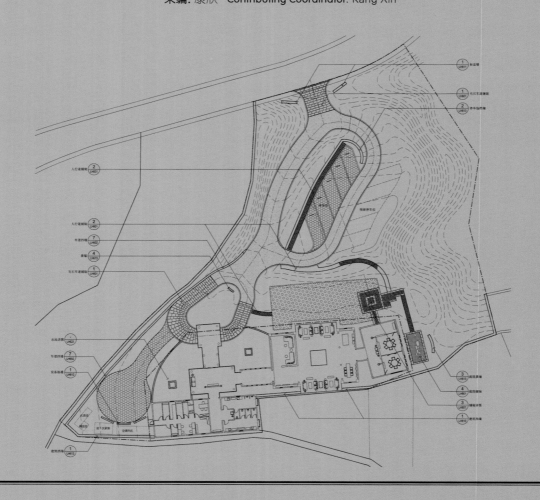

丁力学 总经理
广州市万象设计有限公司

无论是从完整性上、文化上，还是从商业模式、语言上、功能上，这个项目都经得起推敲，并且这个项目还把商业隐含在文化里面，做得比较巧妙。

Ding Lixue General Manager
Visov Design Co., Ltd.

The project is deliberate from integrity, culture, commercial mode, language and function, and it smartly makes business implicit into culture.

洪德成 创始人
DHA洪德成设计顾问（香港）有限公司

这个项目相对来讲比较干净，有主题的延伸，让我看到售楼中心和会所最大的特点——建筑与外部环境、室内关系的沟通和整体呼应。

Hong Decheng Founder
DHA | Dickson Hung Associates Design Consultants

The project is relatively clear, and its themes are extended, which makes the features of the sales center and clubs obviously. The architecture is generally echoing the external and internal environment.

嘉宾点评
Honored Guest Comments

室内设计将山水自然与室内空间充分结合，打造出一个"悠旅山居"的意境。整体设计概念为：朴而不素、华而不奢、雅而不俗，将徽派特有的清雅美学融入现代生活。

设计背景

区域概况

项目选址在全国历史最为悠久的文化古县之一，其中西递、宏村更是皖南古村落的杰出代表。项目步行至西递、宏村景区只需10分钟，出门即享5A级景区的新徽派生活体验。

规划设计

悦榕庄项目集旅游、文化、休闲、居住为一体，在建筑及装潢摆设上均融合当地的特殊风格，以反映当地的风土民情。粉墙、黛瓦、马头墙是众所熟悉的徽派建筑特色，整体设计除了应用这些建筑特色外，色感也延续了传统建筑上的简约稳重。

The interior design fully integrates the natural landscapes and interior spaces, creating a "leisure mountain household" prospect. The overall design concept is pristine, honest and elegant. Hui-style architecture aesthetics are blended in contemporary life.

Designing Backgrounds

Location

The project situates in one of the oldest cultural counties in China, and the Xidi and Hongcun resorts are excellent ancient village representatives in South Anhui province. From the project to Xidi and Hongcun resorts only takes 10 min, and 5A scenic spot is just out of the door for people to experience a new Hui-style living.

Planning Design

Banyan Tree Resort is an all-in-one project of tourist, culture, leisure and residence. The architecture and decoration integrate local style to reflect regional customs. Whitewash walls, black tiles and wharf walls are representatives appeared in Hui-style architecture. Except the application of these architectural features, the project adopts uniform colors to extend the brief and dignified architectural style.

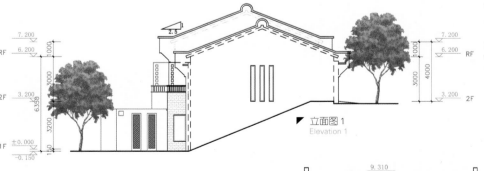

立面图 1
Elevation 1

立面图 2
Elevation 2

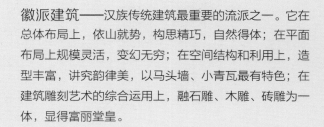

徽派建筑——汉族传统建筑最重要的流派之一。它在总体布局上，依山就势，构思精巧，自然得体；在平面布局上规模灵活，变幻无穷；在空间结构和利用上，造型丰富，讲究韵律美，以马头墙、小青瓦最有特色；在建筑雕刻艺术的综合运用上，融石雕、木雕、砖雕为一体，显得富丽堂皇。

Hui-style Architecture —— is one of the most important ethnic Han architectural styles. It usually lies against hill in overall layout; and enjoys flexible scales in plane layout; and has plentiful forms in spatial structure, especially, it strives for rhythmic aesthetics; wharf wall and gray-green roof tile are its features; it is good at utilizing comprehensive engraving art, such as stone carving, wood carving, brick carving, presenting gorgeous and splendid flavor.

设施配备销售长廊

A. 交汇亭　　I. 服务台
B. 会客室　　J. 会客沙发
C. 走廊　　　K. 壁炉
D. 茶室　　　L. 主模型
E. VIP房　　 M. 地形模型
F. 后勤办公室 N. 休息室 / M
G. 影音室　　O. 休息室 / F
H. 开放式图书室

▶ 展示中心室内一层平面
Exhibition Center Indoor 1F Plan

设计风格

黄山悦榕庄整体设计的概念为：朴而不素、华而不奢、雅而不俗，完整传达徽派特有的清雅美学。室内的设计风格着重于体现徽派中独特的美学，并且融合现代生活的设计观。空间色调整体清亮有泽，个性现代的设计手法，使空间温暖而高雅，完全体现徽派的清雅美感。

设计手法

在设计手法上，采用仿木构的坡屋顶来表现徽派中木构的高脊飞檐，并运用白色大理石雕刻来表现徽派中青石砖雕的精巧艺术成就，更显精致简约。由于整个项目的基地被群山叠抱，所以每个空间都设计了视觉相当穿透的窗景，将自然情趣作为室内空间的视觉延伸，把山水灵气充分带入室内外的氛围中。

家具设计

家具的选择上舍掉了过度的装饰、繁复的细节，以利落的线条来传达山居的简单自然，简洁明快又大气有形，总体上还保留了中式的韵味。设计以明代家具的美感为底，搭配舒适的触感材料，与山水灵气相呼应，自在，静好而不闹。

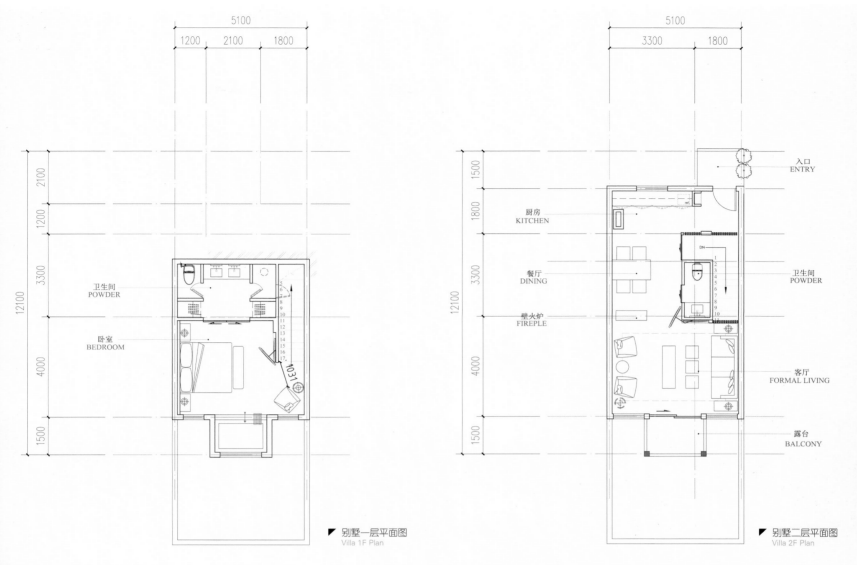

▼ 别墅一层平面图
Villa 1F Plan

▼ 别墅二层平面图
Villa 2F Plan

Design Style
The overall design concept is pristine, honest and elegant, which completely expresses Hui-style architecture aesthetics. The interior design style focuses on Hui-style aesthetics, and integrates modernism style. The space colors are bright and sheen, and personalized modernism methods are used to create a warm and modern space, completely embodying the beauty of Hui-style.

Design Method
The design adopts imitating wooden structures sloping roof to perform Hui-style wooden high ridges and overhanging eaves and white marble engravings to express bluestone brick engravings. The project is surrounded by mountains, so every space is set a transparent visual window views to bring nature to interior space.

Furniture Design
The furniture decorates without excessive adornments or complicated details, but uses neat lines to interpret simple and nature of mountain household. The furniture is in Ming Dynasty Chinese style with high quality materials, echoing the spirit of landscape, liberated and quiet.

游承闵 | Nelson Yu

具有十五年酒店设计经验，曾经在美国、上海及台湾完成过多项五星级酒店项目。作为室内设计部门首席酒店设计师，具有极强的设计及实施能力。

Nelson Yu has 15 years hotel design experiences, and finished many 5-star hotel projects in US, Shanghai and Taiwan respectively. As a Chief Designer in interior design, he is excellent in designing and implementation.

公司简介

HMD注册于英国伦敦，是一间提供多专业工程咨询服务的设计公司。HMD中国的业务范围涵盖了从策划咨询、城市总体规划、城市设计、建筑设计、景观环境设计到室内设计的专业内容。HMD的专家们所追求的是将国际先进的设计理念，高端的建筑科技应用到城市发展中，并使之与中国的市场及传统文化相融合，进而创造出具有独特文化价值及建筑语言的设计作品。

Design Company Profile

HMD is an international multidisciplinary design consultancy, registered in London, UK. The scope of HMD's work in China includes Consultancy, City Planning, Urban Design, Architectural Design, Landscape and Interior Design. What HMD experts are looking for is the application of the advanced international design concepts and the high-end architectural technology in urban development that create a unique design language for their projects and the cultural value.

代表作品

中信青岛胶州璞玉岛，西安华润商场展售中心、融创保利海河大观住宅群、天津融创泰达展售中心、松江会所、浦东皇冠假日酒店，沈阳五星级酒店大堂，AXD Night酒吧，北京私人别墅

Representative Works

Qingdao Jade Island, Xi'an Huarun Shopping Mall Exhibition Center, Sunac Poly Horizon Capital, Tianjin Sunac Taida Exhibition Center, Songjiang Club, Crowne Plaza Hotels & Resorts, Shenyang 5-star hotel lobby, AXD Night Club, Beijing Private Villa

▮ 中信青岛胶州璞玉岛
Qingdao Jade Island

▮ 融创保利海河大观住宅群
Sunac Poly Horizon Capital

▮ 天津融创泰达展售中心
Tianjin Sunac Taida Exhibition Center

▮ 松江会所
Songjiang Club

年度最佳售楼会所空间
THE BEST SALES CENTER SPACE 2014

海南清澜半岛示范区
PENINSULA DEMONSTRATION ZONE, HAINAN

颁奖词
Award Words

尊重自然，崇尚人文的"木屋"

A wooden house, respecting nature and admiring humanity

开发商：丰盛集团　Developer: Full Share Group
项目地址：海南文昌市　Location: Wenchang, Hainan
室内设计：萧氏设计　Interior Design: X.S. Design
软装陈设：LISA LIKAI　Soft Finishing: LISA LIKAI
建筑设计：上海日清建筑设计有限公司　Architecture Design: La Cime INTERNATIONALE PET. Ltd.
项目面积：2 000平方米　Site Area: 2,000 m²
建筑面积：2 855平方米　Building Area: 2,855 m²
设计时间：2013年8月　Design Time: August 2013
竣工时间：2014年2月　Completion Time: February 2014
主要材料：木饰面、芬兰木、棉麻硬包、大理石、透光云石、藤编等
Major Materials: Wood Veneer, Finland Wood, Cotton&Linen Hard Roll, Marble, Transmitting Marble, Plaited Rattan
采编：康欣　Contributing Coordinator: Kang Xin

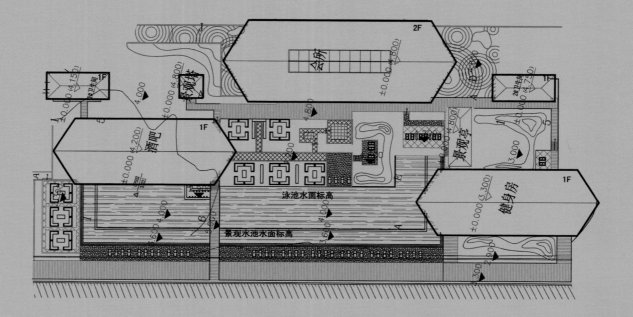

王 冠　主持设计
矩阵纵横

我对海南清澜半岛示范区项目印象比较深，虽然室内设计的部分非常少，但是我们可以感受到建筑师的用心，如果不是两个团队做的话，我们完全可以感受到建筑师对环境的分析，建筑本身的地标性、气势感的感受，包括室内的几个角度也可以看到里面的感觉，同样做得非常精彩。

Wang Guan　Chair Designer
Matrix Design Group

The Paninsula project is impressive to me. Although there is only a little interior design, we still can feel the deep consideration from designers. If the project were not designed by two teams, we could clearly know the environment analysis. It did well in symbolic, imposing manner, as well as several indoor angles.

嘉宾点评
Honored Guest Comments

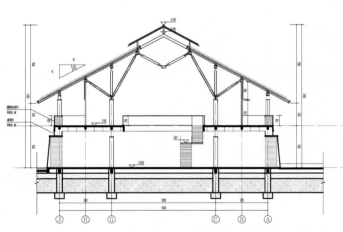
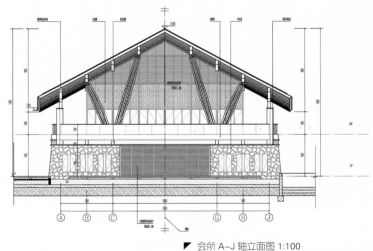

▼ 会所 1-1 剖面图 1:100
　Club 1-1 Section 1:100

▼ 会所 A-J 轴立面图 1:100
　A-J Club Axis Elevation 1:100

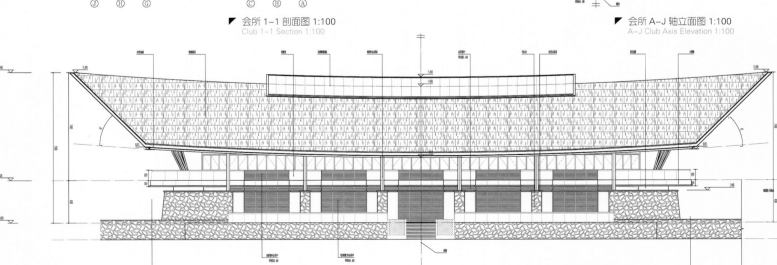

▼ 会所 ⑧-① 轴立面图 1:100
　⑧-① Club Axis Elevation 1:100

屋面大而陡峭的"木屋子"和谐地与三面海景融为一体，合理的功能分区布局，天然茅草和生态木与室内结构体系的搭配，使传统与时尚相融合，创造出一个休闲度假的空间，传达出积极的场所意义。

设计背景

区域概况

海南清澜半岛位于海南文昌市高隆湾北端，为270°双湾三面海景的人工岛屿，毗邻海南卫星发射基地。该示范区在半岛西面沿海处，主要业态包括一幢两层的会所、一幢一层的健身中心及一幢一层的酒吧。

项目规划

项目前期定位为以打造休闲活力湾区为目的的高端旅游地产项目，以度假人群为主。项目周边配套成熟，拥有高隆湾、八门湾、红树林等丰富的旅游资源。邻近中国最大的火箭发射基地，同时也是中国唯一可供观光的火箭发射基地。

建筑设计

示范区的建筑设计风格试图体现典型的热带雨林环境和传统的民族风情特色。屋面大而陡峭，采用自然茅草铺盖，四角为石材基座，以稳重的造型塑造建筑。三幢建筑由会所前的水系串联，以水系方式联系若干个不同特点的庭院，创造休闲、怡情的浪漫空间。

园林设计

示范区的园林设计以代表性的热带植物为造景元素，创造出密林、疏林、林带、草地、花境等颜色、形态各异的立体景观。大到园林空间的打造，小到亭台水榭的细节装饰，都体现出对自然的尊重以及对人文的崇尚。

The wooden house has a large and deep roof, which harmoniously integrates with seascapes at three sides. Proportional function layout, natural thatch and wood echo indoor structure system. It is a good leisure resort, conveying positive significance.

Designing Backgrounds

Location

The Peninsula project stands at the north of Gaolong Bay, Wenchang, Hainan, where is an artificial island with 270° two bays and three-seascapes, and is close to Hainan Satellite Launch Base. The demonstration zone is beside the west of the island, including a two-layer club, a one-layer fitness center and a one-layer pub.

Planning

The project is positioned on forming a leisure bay high-end tourist resort, catering for tourists. It has mature surrounding supporting facilities and plentiful tourist resources, including Gaolong Bay, Bamen Bay and Mangrove Forest. It is close to the biggest satellite launch base in China; meanwhile, it is the only one can be visited.

Architecture Design

The design aims at presenting the environment of typical tropical rainforest and traditional ethnic customs characteristics. The large and steep roof, natural thatch and stone pedestals at four corners form a calm and steady architecture figure. There are three houses connected by water system, every house has its own typical yards, relaxing and romantic.

Garden Design

The garden design takes tropical plants as landscaping elements, creating stereoscopic landscapes in different colors and forms, including dense forests, open forests, forest belts, grasslands and flower beds. Every detail shows respects to nature and humanity, no matter in garden space or in pavilion decorations.

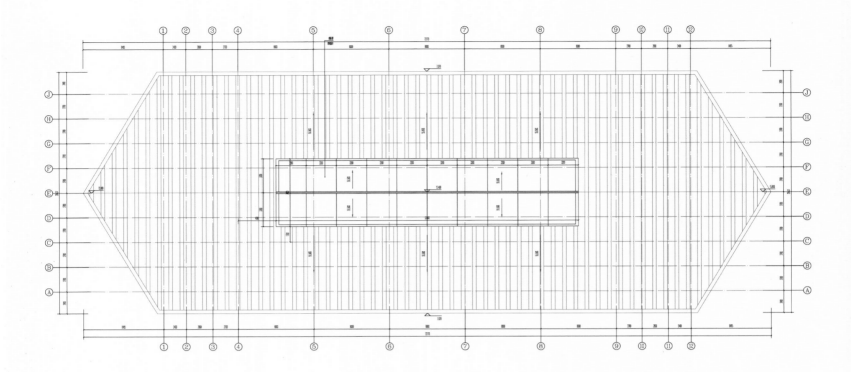

▶ 屋顶平面图 1:100
Roof Plan 1:100

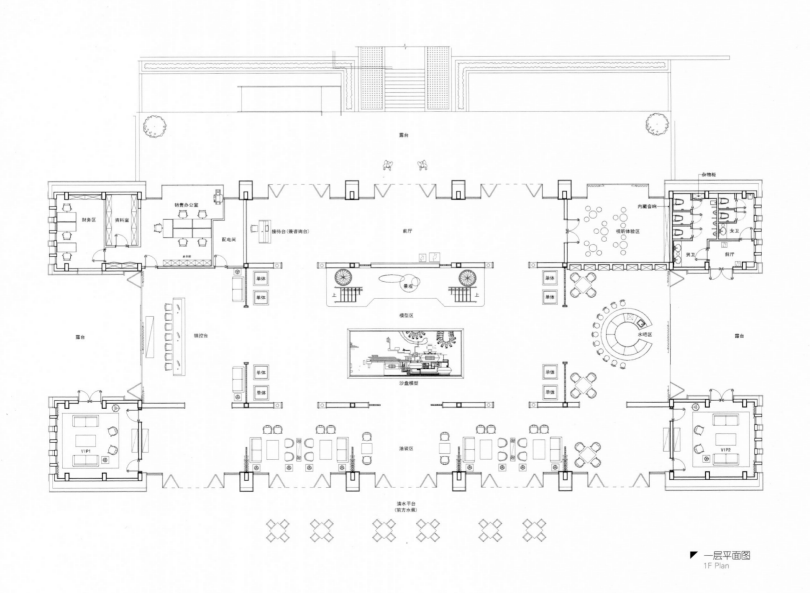

▶ 一层平面图
1F Plan

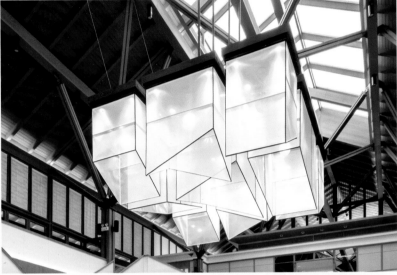

功能布局

朝南的一线海景，与入口的长廊形成对称，成为"洽谈区"。当客人经过入口长廊进入沙盘区后，销控台即在眼前。合理的功能分区布局，优美的海景，舒适的沙发和座椅，让客人愿意长久逗留，刺激购买欲望。

会所大堂

会所大堂内环绕中庭四面的大楼梯将上下两层空间贯通，正对大海，视野开阔。结构体系采用钢结构的分叉支撑起一个三维大悬挑的弧面屋顶，并将梁柱完全暴露，使结构的美感和纤细的支撑构件与"木屋子"合为一体，创造出层叠、变幻的韵律。同时，通过材料与结构的轻盈使传统与时尚体现出隐含的融合，使场所的地域性得到进一步的表达和提升。

Function Layout

The entrance gallery is facing the south seascapes, becoming a "negotiation zone". Once the customers go through the gallery into sand table area, the reception desk is witnessed. The proportional function layout, beautiful seascape, cozy sofas and chairs make customers linger on in the sales center and have strong purchasing desire involuntarily.

The Club Hall

There is a grand staircase around the atrium of the club hall connecting the upper and lower two layers of spaces. The club is facing the sea with broad vision, and it adopts steel structure to hold up a three-dimension arc roof. The beam columns are exposed completely, making the beauty of structure and slender holding components unite with the "wooden house" seamlessly. Meanwhile, the lithe materials and structures make tradition and fashion mix together implicitly.

透光云石

一种新型的复合材料，由高分子材料合制而成。透光云石的份量轻（与天然石材相比）、硬度高、耐油耐脏耐腐蚀。板材具有不变形，防火抗老化，无辐射、抗渗透等特点，可根据客户的需求随意弯曲，无缝黏接，真正达到浑然天成的境界。

Transmitting Marble

A new compound material, is made of high polymer materials. It is lighter than other natural stones but with higher hardness and also oil-proof, resisting staining and anti-corrosion. It has other advantages, such as fire-proofing, anti-aging, non-radiative, impervious, and it also can be bended freely and seamless bond.

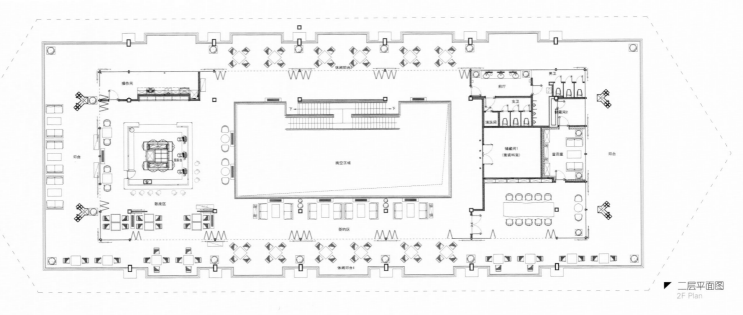

▼ 二层平面图
2F Plan

私密与开放兼备

为了保证私密性,附属用房被单独设置到了三幢建筑之外,可通过穿行路线到达;会议室等空间均被设置于建筑四角的基座之内。因此建筑内部几近全开敞,创造出休闲、开放的交流空间。而设备均采用地面暗藏的方式以突出细部的整洁。

特色材质

海南沿海地区会受到海风及海水的侵蚀,要求材料具有耐候性、耐酸性和耐腐蚀性并且还能够抵抗台风,因此设计师选用了相比人工材料更加具有优势的天然茅草和生态木,在新材料的基础上进行了构造节点的创新与改造,体现结构之美、自然之美。

Privacy and Openness

Affiliated rooms are set outside of the three houses to ensure privacy, which are accessible by travel routes. Conference rooms are set in pedestals at four corners, so the interior space is un-closed, creating a leisure and open communication space. All the facilities are concealed underground to highlight neat details.

Featured Materials

Hainan is close to coast land where often suffers the erosion of wind and seawater, so architecture materials ask for acid-resisting, corrosion-resisting and resistant ability for typhoon. Designers use natural thatch and ecological wood which are better than artificial materials, and innovate and transform tectonic nodes on the basis of new materials, expressing the beauty of structure and beauty in naturalness.

年度最佳售楼会所空间
THE BEST SALES CENTER SPACE 2014

广州萝岗绿地中央广场售楼处
GUANGZHOU GREENLAND CENTRAL PLAZA SALES CENTER

颁奖词
Award Words

独特的空间一体化设计、错层式功能布置，非常具有时代代表性
Unique space integrated design and split-level function layout boast representativeness of the times

开发商：广东绿地事业部　Developer: Greenland Group Guangdong Business Division
项目地点：广州市萝岗区科学大道与神舟路交汇处
Location: The intersection of Science Avenue and Shenzhou Road, Luogang District, Guangzhou
设计公司：JWDA骏地设计　Design Company: JWDA
设计时间：2013年7月　Design Time: July 2013
竣工时间：2013年12月　Completion Time: December 2013
建筑面积：2 000平方米　Building Area: 2,000 m²
主要材料：穿孔板、锌板　Major Materials: Perforated Plate, Zinc Plate
采编：谭杰　Contributing Coordinator: Tan Jie

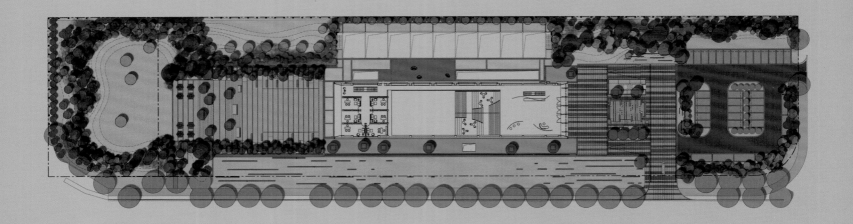

李仕泉 副总经理
金盘地产传媒有限公司

这个项目做得非常专业，设计师注重整体形象的打造，克服了狭长空间设计的局限，给人带来不一般的视觉体验。

Li Shiquan Vice-general Manager
Kinpan Estate Media Ltd.

It is a professional work. Architects lay great emphasis on overall image and conquer the limit of long yet narrow space, bringing extraordinary visual experience.

嘉宾点评
Honored Guest Comments

空间一体化的推敲，打破了传统售楼中心平层效果，以大台阶的走入式感受吸引客户，带来不一样的感受。设计以"城市广场"为主线，通过错层式功能布置，流线的合理分流使得不同性质的客户各得其所。

设计背景

区域概况

萝岗绿地中央广场距离广州中心区28公里，占地10万平方米，总建筑面积约60万平方米，将被打造成广州城市东进门户的60万平方米商务综合体地标。首期产品包括1栋100米高标准甲级写字楼、2栋LOFT复式公寓、1栋酒店式公寓及部分欧美时尚BLOCK街区商铺。

设计理念

该售楼处是项目住宅部分的销售中心。为了满足业主方的时间及功能需求，设计本着快速建造及一体化设计的原则，将建筑、景观、室内设计同步进行。

建筑设计

设计师将"山水""雪花"主题融入在建筑和景观设计中，建筑表皮以形意化的方式还原"羊城八景"之一的"萝岗香雪"，而简洁的建筑体型又具备强烈的标志性，景观设计表现出山水意境。

The integrated space breaks traditional sales center flat bed staleness, and grand steps naturally absorb clients in. The design takes "urban square" as a mainline and uses split-level function layout and rational pedestrian circulation to meet various demands.

Designing Backgrounds

Location

There is 28 km from Guangzhou center. It occupies 100,000 m² site area and has 600,000 m² gross building area. The Guangzhou Greenland Central Plaza Sales Center would be the commercial complex landmark in Guangzhou city east portal. The first phase products include one high grade-A office building of 100 m, two LOFT apartment buildings, one service apartment building and some fashionable BLOCK shops in European and American style.

Design Concept

It is the residential sales center of the project. In order to meet the time and functional requests of clients, the design complies with the principles of rapid construction and integrated design, and simultaneously implements architecture, landscape and interior decoration.

Architecture Design

The themes of "Shanshui" and "snowflake" are blended in architecture and landscape design. The facade uses an ideographic format to make a revivification of "Luogang Xiangxue", one of the "Eight Sights of Guangzhou". The concise building shape is intensely iconic, while the landscape presents Shanshui artistic conception.

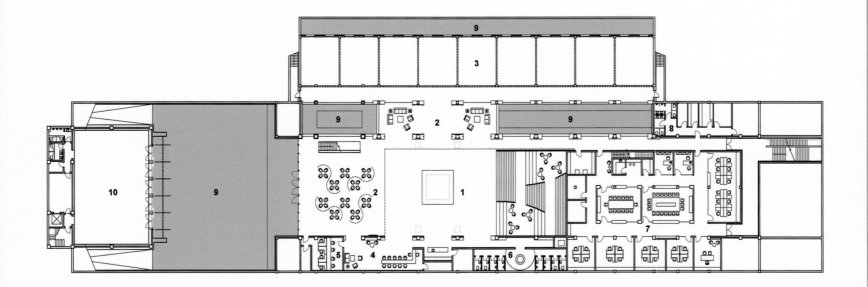

1. 楼盘模型展示区
2. 休息区
3. 样板间
4. 银行业务咨询处
5. 财务室
6. 洗手间
7. 办公室
8. 设备房
9. 花园
10. 宴会厅

▼ 地下层平面图
　 Basement Plan

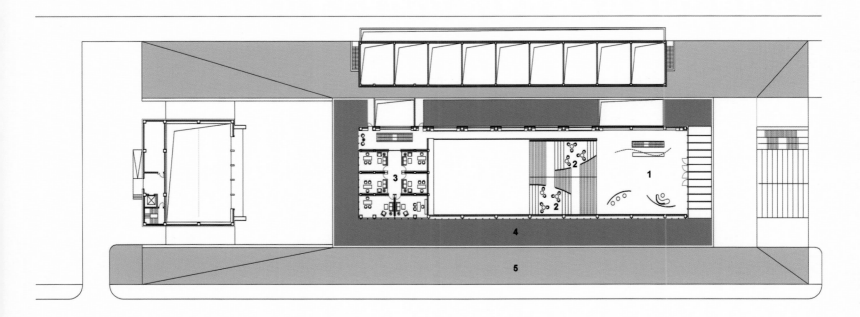

1. 接待大堂
2. 休息区
3. VIP 房
4. 水景
5. 花园

▼ 一层平面图
　 1 F Plan

空间设计：一体化

设计对空间一体化的推敲打破了传统售楼中心的平层效果，以大台阶的走入式感受吸引客户，传统的洽谈室也以大屏幕的方式存在，让顾客不管是刚进入到销售中心还是在看沙盘，亦或在洽谈中，都能沉浸在该项目的氛围中。尽管身处室内，仍会给销售过程带来不一般的视觉体验之旅。

功能区域

该项目的用地范围较为狭长，为突破这一局限，设计以"城市广场"为主线，组织销售展示中心各部分功能。各功能区域之间通过台阶及坡道联系，形成不同高度的展示空间，为起伏变化的"城市广场"提供丰富的空间体验，又通过宽阔的下沉广场及庭院采光，消除了半地下空间的闭塞感。

错层设计

设计通过错层式功能布置，组织不同的功能流线，如到达流线、参观流线、服务流线、离开流线均相对独立，避免交叉，为销售行为提供最大的宽容度。错层设计使得流线布局更加合理，让不同性质的客户各得其所，避免了流线混杂的问题。

Space Design: Integration

The integrated space breaks traditional sales center flat bed staleness, and grand steps naturally absorb clients in. Traditional reception room is added with a large screen, so no matter watching sand table or negotiating, people would immerse in the project atmosphere. Although indoor, the sales experience would be extraordinary.

Functional Region

In order to break through the limit of long and narrow landform, the design takes "urban square" as mainline to arrange each part function of the sales center. Steps and gentle slops are used to connect different functional parts, forming different heights of display spaces and bringing plenty of spatial experience in the fluctuant "urban square". The broad sunken square and courtyard light setting remove the sense of occlusion in the half underground space.

Split-level Design

The design employs split-level layout to organize functional circulation, such as arriving, visiting, serving and leaving circulations which are independent and avoided crossing. The arrangement leaves the biggest tolerance for selling.

品质

商业空间在消费潮流的影响下,也在向更高层次发展。商业空间的品质,不仅在于新的方法和材料的应用,商业机能性与环境塑造的提升,更在于设计语言与空间形态、科技感与时尚的完美结合,催生购买欲并使消费者对购买的商品产生信任感。

艺术

商业空间里面创造出的生活场所感和文化艺术感是非常重要的。其讲求的是空间体验性。针对目标受众,在艺术的情境设计上有趣味、创新和引领性,做到既能吸引客人,又尊重商品与人,更能引领新的生活方式和消费方式。

人居

好的空间规划、好的业态组合和商业人流动线整体营造一个好的购物体验,再加上众多休闲娱乐设施,创造一个综合的空间场所。即使消费者不去买东西,进入里面也能做其他事情,得到精神上的满足与享受。

价值

商业空间的价值不再仅仅局限于设计,设计者与开发商意在打造有主题、有价值的空间。从产品角度来讲,它是一个优秀的设计;从消费者角度来讲,它能让消费者逛得舒服,享受到满意的服务。

Quality

Commercial space has been developing for a higher level under the consumption trend influence. The application of new methods and materials, commercial functions and environment improvement have been acting on commercial space quality, while design language, spatial forms and combination of technology and fashion will further work on the quality, which is a better way to stimulate purchase desire and obtain consumers' trust for commodities.

Art

To create life arena and cultural artistic feeling in commercial space is crucial, which is particular in spatial experience. According to intended consumers, funny, innovative and steering artistic scenario design is created so as to absorb consumers and at the same time show respects to people and commodities, furthermore, to lead life-style and pattern of consumption.

Habitat

A good shopping experience depends on great spatial planning, formats combination and commercial pedestrian circulation, and when leisure entertainment facilities are added, a comprehensive space will be created. In this way, consumers go in there, even if they do not buy anything, they receive spiritual satisfaction and enjoyment still.

Value

The value of commercial space not only lies in design, but in the themes and valuable spaces generated by designers and developers. A good commercial space firstly is an excellent design from product view; secondly, it is an awesome shopping destination for consumers and meanwhile they get satisfying service.

随着经济的发展，商业空间的操作日益成熟，但平面功能也日益倾向于模式化，能在符合商业的条件下做出创新就显得尤为难得。金盘奖活动嘉宾认为，当下的商业空间设计更应该注重消费者的体验性，在购物中心的公共区域创造出生活场所感，塑造出有价值感的商业空间。如果设计者加入太多自己的想法进去，反而会减弱空间的品质。

With the development of economy, commercial space operation mode is increasingly mature, nevertheless, plane functions are gradually stereotyped. Therefore, innovation under proper commercial conditions is praiseworthy. Kinpan Award participants hold that commercial space should take consideration of consumers' shopping experience, and create life arena at public zone so as to form a valuable commercial space. If designers add their own extra ideas, the space quality would be decreased.

年度最佳商业空间
The Best Commercial Space 2014

湖州东吴银泰城
Dongwu Intime City, Huzhou

郑州锦艺城购物中心
Jinyi City Shopping Mall, Zhengzhou

大连高新万达广场室内步行街
Wanda Plaza Indoor Pedestrian Street, Dalian

年度最佳商业空间
THE BEST COMMERCIAL SPACE 2014

湖州东吴银泰城
DONGWU INTIME CITY, HUZHOU

颁奖词
Award Words

立足于小城市的消费心理需求，丰富美陈装置营造活泼商业氛围

Basing on consumer psychology in small city, rich art decoration creates a lively commercial atmosphere

开发商： 湖州银东购物中心有限公司　**Developer:** Huzhou Yindong Shopping Mall Co., Ltd
项目地址： 浙江省湖州市吴兴区劳动路518号　**Location:** No.518 Laodong Road, Wuxing District, Huzhou, Zhejiang
室内设计： 北京清尚建筑设计研究院曾卫平商业设计
Interior Design: Beijing QingShang Landscape Architecture Design Institute ZWP Commercial Design
主创设计师： 曾卫平　**Chief Designer:** Zeng Weiping
占地面积： 49 976平方米　**Site Area:** 49,976 m²
建筑面积： 102 500平方米　**Building Area:** 102,500 m²
竣工时间： 2014年5月　**Completion Time:** May 2014
采编： 康昭生　**Contributing Coordinator:** Kang Zhaosheng

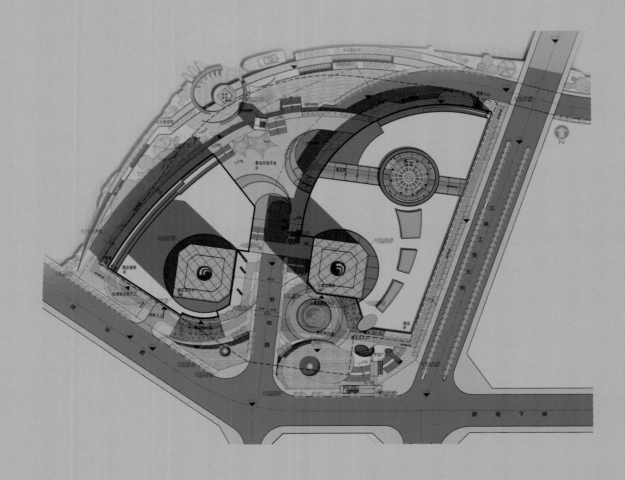

中国建筑新闻网

设计是为商业运营服务的。在这个项目中，我们根据业态进行定制化美陈设计，为商业空间添彩，正是对这一理念的实践。

NewsCCN.com

Design services for business operation. In this project, we customized art decoration design according to retailing format to enrich the commercial space.

嘉 宾 点 评
Honored Guest Comments

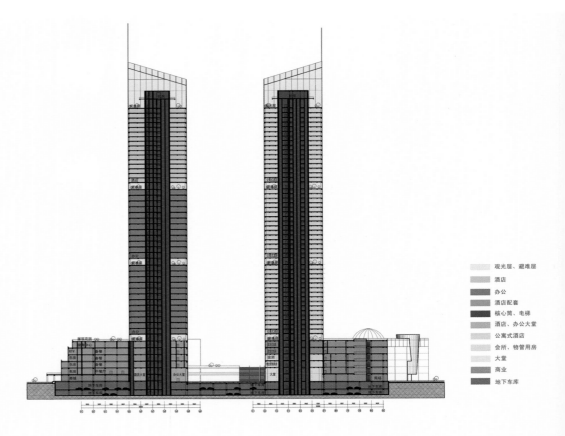

空间设计着眼于对商业氛围的提升，从湖州当地人群的消费心理出发，打造出丰富多变的空间效果。大量富有趣味性的美陈装置活跃了空间氛围，凸显了不同业态的主题，与消费者建立起互动，为消费者提供了丰富而独特的休闲购物体验。

设计背景

项目概况

本案是东吴国际广场的一部分，广场由两栋288米高的大楼组成，内设五星级酒店、甲级写字楼、空中国际公馆、酒店式公寓、高层主题观光旅游和银泰城；银泰城的最初定位是大型艺术旅游购物中心，是两座高塔的底商。整体商业面积约10万平方米，地上六层，地下一层，每一层都有不同的主题体验。

建筑风格

银泰城的建筑设计简洁大气，色调典雅沉稳，主要造型语言为直线与方形块面的组合与切割。空间设计考虑到这种现代感十足的建筑基调，将空间设计语言与建筑语言完美结合，打造简洁大气、典雅沉稳的空间氛围。

Space design aims to enhance commercial atmosphere. Basing on consumer psychology of local people, the project creates a rich and varied spatial effect. Numerous funny art decorations activate atmosphere, highlight themes of different commercial activities and build interaction with consumers, thus, provide consumers with a rich and unique shopping experience.

Designing Backgrounds

Overview

This project is a part of Dongwu International Plaza. The plaza includes two 288 m high towers, in which are a 5-star hotel, a grade-A office building, air international mansions, serviced apartments, high-rise theme sightseeing and the Intime City. Intime City, located at the bottom of the two towers, is initially positioned as a large art tourist shopping center. The gross commercial area is about 100,000 m² covering the 1 to 6 floors on ground and 1 underground floor. Each floor has its own theme.

Architectural Style

The architecture design of the project is concise and graceful, its color is elegant and dignified and its main shape languages are compositions and segments of straight lines and rectangles. In consideration of the modern architectural style, the space design perfectly combines space design language with architectural language to create a concise, grand, elegant and dignified spatial atmosphere.

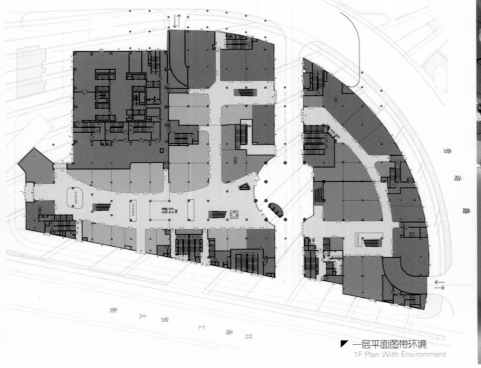

▶ 一层平面图带环境
1F Plan With Environment

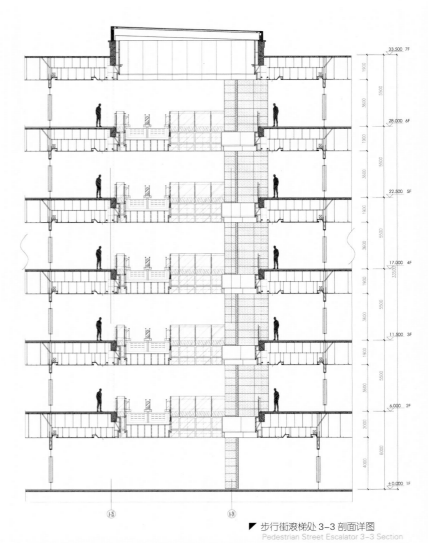

▶ 步行街滚梯处 3-3 剖面详图
Pedestrian Street Escalator 3-3 Section

商业布局

银泰城在体量上比一般的百货商场大，比大型购物中心稍小，一、二层为快时尚，包括时尚服饰和餐饮；三层主要为婴幼儿服饰用品；四层以KTV、电玩城、运动服饰为主；五层主打影院，兼有餐饮；六层以餐饮为主。

设计思路

设计团队将空间设计如何助推商业的更大成功作为设计的核心切入点，针对湖州作为三线城市的地域及文化特点，围绕业态及运营展开，力求通过设计营造独特的体验式购物环境。湖州不是一二线城市，当地高端消费群体的消费心理和喜好与一线城市不一样，人们更喜欢丰富的色彩与多变的空间，所以设计运用了大量的美陈装置，力求营造丰富的空间感受。

Commercial Layout

Volume of the project is larger than a common department store, yet smaller than a large shopping center. Its first two floors are for fast fashion, including fashion clothes and F&B; its 3rd floor is for baby clothes and stuffs; its 4th floor is KTV, video games, sportswear shops; the 5th floor features cinema combining restaurants; and the 6th floor is mainly restaurants.

Design Idea

The design team tries to promote commercial activities by space design. The design is unfolded according to retailing format and operation on the basis of the local culture to create a unique experience shopping environment. Huzhou is a third-tier city and the local high-end consumers have different consumption psychology from people in first- or second-tier cities. People here prefers rich colors and varied spaces, therefore, the design team applies a large amount of art decoration to create a colorful space feeling.

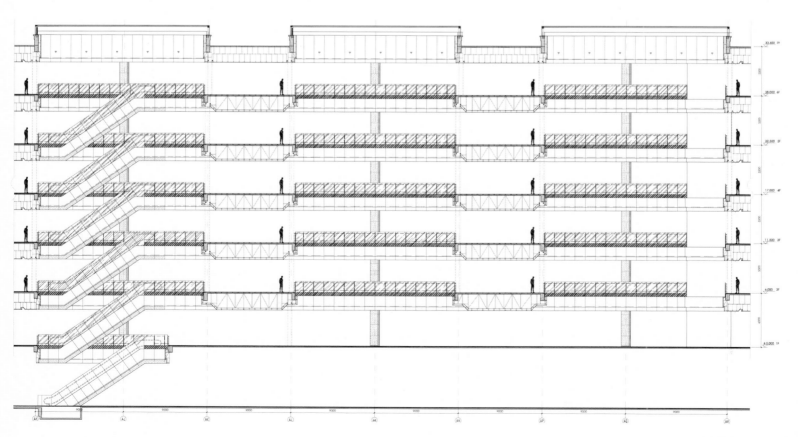

▶ 步行街 1-1 展开剖面图
Pedestrian 1-1 Profile

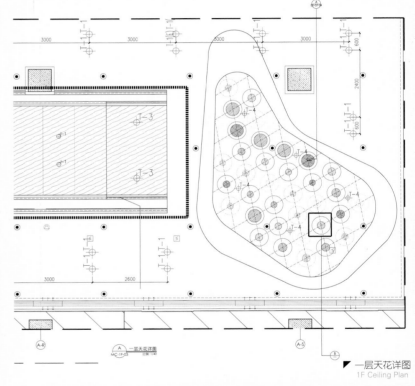

▼ 一层天花详图
1F Ceiling Plan

美陈设计——美陈是现代新兴起的学科，是集广告专业、室内专业、环境景观艺术、工业产品设计等设计门类的综合性很强的专业，它与一般广告不同，更强调突出造型的艺术主题、表现手法等。美陈一般分为商业美陈与公共美陈。商业美陈是以环境和装饰艺术为主体，结合商业建筑、商业运营、营销策划的综合性项目。

Art decoration design — it is a burgeoning subject, a domain of strong comprehensiveness, assembling advertising, interior design, environmental landscape, industrial product design, etc. Different from common advertisement, it stresses art theme and technique of expression. It is generally divided into commercial art decoration and public art decoration. The commercial one takes environment and decorative art as main body and combines commercial building, commercial operation and marketing plan as a whole.

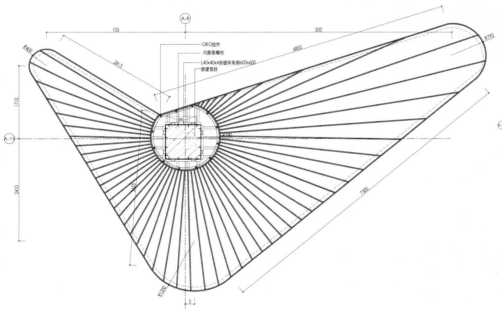

▼ 四层柱子横剖节点详图
4F Cross Section Node

▼ 五层天花详图
5F Ceiling Plan

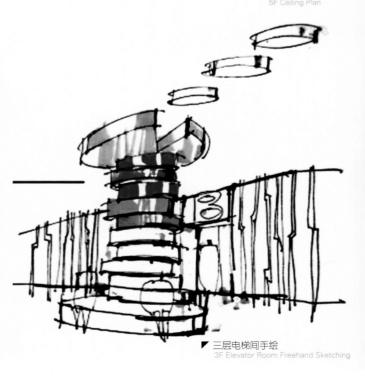

▼ 三层电梯间手绘
3F Elevator Room Freehand Sketching

丰富的美陈设计

室内设计根据湖州银泰城的业态,重点在美陈设计上做文章。美陈设计使业态主题更明确直观,既可活跃空间氛围,也可以随着未来商业业态的调整而灵活变化。

购物中心主入口外的动感彩色雕塑丰富了空间层次,将室外的人流吸引过来;扶梯和电梯厅的入口处设置玻璃钢装置,将咨询台和休息功能相结合,体现空间的穿插关系;室内长街的中庭通过地面拼花和模数尺寸的变化构成棋盘,并放置国际象棋,人走在其中有很强的参与感;二层不锈钢材质的异形柱体,与柔软的鹅卵石造型座椅形成刚与柔、亮与暗的对比;三层厅内的绿色、红色、白色等亮色系柱子采用了可爱的卡通造型;四层设置不锈钢材质的音乐主题雕塑,为人们进入KTV充当前奏,天花上的悬挂光碟和发光音符是该区域的美陈要点;五层采用放映机胶片转盘作为墙面装饰,富有趣味性。

Luxuriant Art Decoration

According to the commercial activities of the project, its interior design focuses on art decoration, which not only activates atmosphere, but also flexibly changes with the adjustment of future retailing format and clears the commercial theme.

The dynamic colored sculpture in front of main entrance enriches the space layers and attracts visitors. Entrances of escalators and elevators are equipped with fiberglass, combining functions of inquiry desk and lounge to present an alternation of spaces. In atrium of indoor street, parquet flooring of varied modular sizes constitutes a chessboard on which chesses are placed. People will feel a strong sense of participation when walking through the atrium. Columns of irregular shapes and stainless steel on the 2nd floor form tough-tender and bright-dark comparisons with chairs of gentle pebble shapes. Columns of bright colors on the 3rd floor adopt lovely cartoon images. Music theme sculptures of stainless steel on the 4th floor serve as prelude as people entering the KTV, and the hanging disks and shining notes act as key points of art decoration here. The wall decoration of projector film turntables on the 5th floor is funny.

年度最佳商业空间
THE BEST COMMERCIAL SPACE 2014
郑州锦艺城购物中心
JINYI CITY SHOPPING MALL, ZHENGZHOU

颁奖词
Award Words

简洁设计打造华丽视觉效果，大气空间释放流畅时尚律动
Concise design creates luxury visual effect, graceful space diffuses fluent and vogue rhythm

项目地址：河南省郑州市中原区　**Location:** Zhongyuan District, Zhengzhou, Henan
设计公司：J&A姜峰设计公司　**Design Company:** Jiang & Associates Design Co.,Ltd.
建筑面积：165 000平方米　**Building Area:** 165,000 m²
主要材料：天然石材、铝板、玻璃、人造石、树脂板、不锈钢、波浪板和木饰面等
Major Materials: Natural Stone, Aluminum Plate, Glass, Artificial Stone, Resin Plate, Stainless Steel, Waved Plate, Wood Veneer
采编：陈惠慧　**Contributing Coordinator:** Chen Huihui

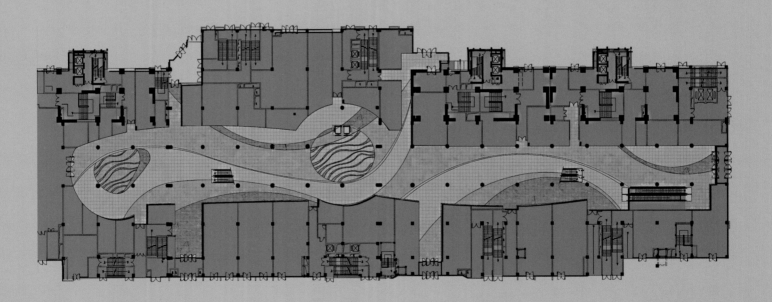

新浪网
锦艺城购物中心空间内部结构宽阔通透，中庭挑空自然采光，同时设计更充分彰显时尚魅力，让人购物的同时享受视觉之美。

Sina.com
Jinyi City Shopping Mall has a spacious and transparent interior structure and hollowed atrium of natural lighting. Besides, this design manifests a vogue charm, which enables people to enjoy the visual beauty when shopping.

嘉　宾　点　评
Honored Guest Comments

设计背景

项目背景

该购物中心是锦艺城项目的一部分。锦艺城是总规模达120万平方米的综合项目，由高品质住宅锦艺国际华都、30万平方米时尚购物中心锦艺城购物中心、西区商务地标写字楼锦艺国际中心以及小学、幼儿园、综合性会所等组成。高端的住宅组团将为商业项目带来稳定的高收入人群。

配套设置

锦艺城购物中心是郑州市规模最大的单体购物中心，由锦艺集团斥资50亿打造，项目设置1800个停车位，70余部电梯，豪华中庭设计，多个空中连廊、观光过道……这些都将提高消费者的购物兴趣。

Designing Backgrounds

Project Backgrounds

The shopping mall is a part of Jinyi City. Jinyi City, a large complex of 1,200,000 m², is composed of a high-quality residence Jinyi Classic Life, a 300,000 m² Jinyi City Shopping Mall, a landmark office building Jinyi International Center, a primary school, a kindergarten and a club. The high-end residential cluster will bring the shopping mall stable consumers with high income.

Supporting Facilities

The shopping mall is the largest shopping center in Zhengzhou, which has cost Jinyi Group 5,000,000,000 RMB. It is equipped with 1,800 parking spaces, over 70 elevators, a luxury atrium, several air corridors and sightseeing galleries… All these boost consumers' shopping mood.

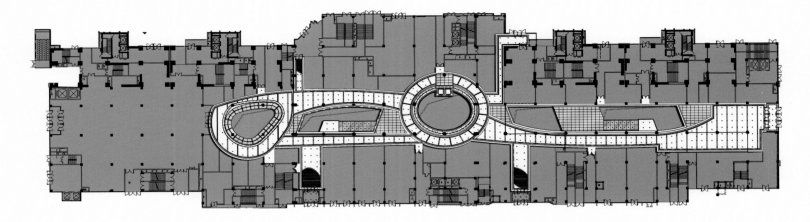

▶ 一层天花平面图
1F Ceiling Plan

▶ 一层地材平面图
1F Flooring Plan

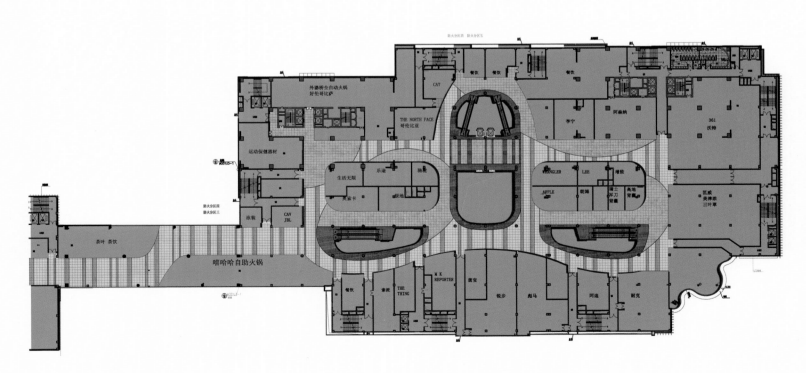

▶ 三层地材平面图
3F Flooring Plan

商业构成

项目营业楼层为地下1层至地上4层，总商业面积约30万平方米。其中，大型购物中心、精品百货、综合超市、专卖店、文化娱乐、餐饮、商务等业态能够满足不同消费群体需要；业态组合中40%是零售、20%是餐饮、20%是休闲娱乐，还有20%是大型运动、家居建材、数码产品等，提倡一种全新的国际化生活方式。

设计思路："锦"的质感

购物中心由两栋建筑组成，定位于温馨家庭和时尚青年两类消费群体，并以此进行分区处理。设计遵循在装饰上整体统一、区域强调变化的原则，将外部建筑的"锦"的理念延续到室内设计中来，强调"锦"的质感，形成呼应，达到里外均衡的效果。

空间设计：内外呼应

空间设计整体风格呈现出柔顺、华丽、流畅的特点，展现出高雅、时尚、富有韵律的动感空间。中庭设计简洁大气，色调明朗。设计将室内外景观、小品与商场主题密切联系，保证内外景观、空间及主题特色的和谐统一。玻璃采光顶以及环形设计，形式新颖，在满足采光的前提下，与室内、室外的整体设计相融合。

Commercial Composition

The retail part of the project covers underground 1st floor to ground 4th floor of 300,000 m² gross commercial area. In this part, the large shopping center, boutique department store, supermarket, exclusive store, culture entertainment, restaurant, business, etc., satisfy demands of different consumers. The commercial activities contain 40% retail, 20% F&B, 20% leisure entertainment and 20% sports, home building materials, digital products, etc., advocating a brand-new and international life mode.

Designing Idea: Texture of "Brocade"

The shopping mall is composed of two buildings which respectively target warm families and fashion youths. Its design follows the principle of unified decoration and varied zones. Besides, it continues the concept of "brocade" of exterior design into the interior design, emphasizing the texture of brocade and achieving an internal and external equilibrium.

Space Design: Inside and Outside Echoing Each Other

The general style of the space design presents a gentle, luxury and fluent feature and unfolds an elegant, vogue and dynamic space. The design of atrium is concise and graceful with bright colors. Indoor and outdoor landscapes, featured landscape and theme of the mall are connected tightly to ensure a harmonious unification of landscapes, space and theme. Glass lighting ceiling adopts a ring design, which is novel and integrates with the general design on the premise of sufficient lighting.

姜峰 | Frank Jiang

J&A姜峰设计公司创始人、董事长、总设计师。现担任中国建筑学会室内设计分会副理事长、中国建筑装饰协会设计委副主任、中国室内装饰协会设计委副主任等社会职务。

Founder of Jiang &Associates Design, Chairman, General Designer。 Deputy Director of China Institute of Interior Design, Deputy Director of China Building Decoration Association Design Committee, Deputy Director of China National Interior Decoration Association Design Committee.

所获荣誉 Awards & Honor

亚太十大领衔酒店设计人物

终身艺术设计成就奖

中国室内设计功勋奖

中国酒店设计领军人物

深圳市十大杰出青年

深圳百名行业领军人物等社会荣誉

并入选美国《室内设计》杂志名人堂等

Top10 Asian Leading Hotel Designers

"China Exterior Design 20 Years" Design Meritorious Prize

China Interior Design Exploit Award

The Leader of China Hotel Design

Honorary Title of Shenzhen Ten Outstanding Young Persons

Honorary Title of Shenzhen 100 Industry Leaders

Selected Into "Hall Of Fame" of INTERIOR DESIGN Magazine, etc.

公司简介

J&A姜峰设计有限公司，简称J&A，是由荣获国务院特殊津贴专家、教授级高级建筑师姜峰先生及其合伙人于1999年共同创立。现拥有来自不同文化和学术背景的四百人的国际设计团队，经过十五年的发展，J&A沉淀了属于自己的设计理念——做最好的设计，这并非是一句空洞的口号，可以概括为四个关键词：未来、理想、价值、完美。

Design Company Profile

Jiang & Associates Design Co.,Ltd. (J&A), was found by Frank Jiang and copartners. Mr. Jiang is an expert in Allowance of State Department and a senior architect of professor level in the same time. The company has about 400 designers with different cultural and academic backgrounds, and after 15 years' development, J&A deposits its own design concept—to do the best design, which is not an empty slogan but a summary of four words: future, ideal, value and perfection.

代表作品

上海浦东文华东方酒店、大连国际会议中心、深圳市市民中心、深圳会议展览中心、上海华润五彩城、西安华润万象城、深圳星河时代COCO Park、深圳宝能all city购物中心、郑州西元国际广场、龙湖时代天街、深圳四季酒店、天津圣瑞吉酒店、深圳丽思·卡尔顿酒店、深圳地铁车站、大连文化中心等

Representative Works

Mandarin Oriental in Pudong Shanghai, Dalian International Conference Center, Shenzhen Civic Center, Shenzhen Convention and Exhibition Center, Dreamport in Shanghai, The MixC in Xi'an, Shenzhen Galaxy Times COCO Park, Shenzhen All City, A.D.Plaza, Longfor Times Paradise Walk, Four Seasons Hotel Shenzhen, ST.Regis Tianjin, The Ritz-Carlton Shenzhen, Shenzhen Subway Station, Dalian Cultural Center

▼ 上海浦东文华东方酒店
Mandarin Oriental in Pudong Shanghai

 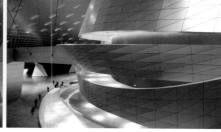

▼ 大连国际会议中心
Dalian International Conference Center

▼ 深圳星河时代 COCO Park
Shenzhen Galaxy Times COCO Park

▼ 深圳宝能 all city 购物中心
Shenzhen All City

年度最佳商业空间
THE BEST COMMERCIAL SPACE 2014

大连高新万达广场室内步行街
WANDA PLAZA INDOOR PEDESTRIAN STREET, DALIAN

颁奖词
Award Words

多种设计手法穿插，细节把控合理，树立了A级店内装品质的新标杆
Multiple design methods and rational grasping details establish a new benchmark of grade-A interior decoration quality

开发商: 大连高新万达广场投资有限公司　**Developer**: Dalian High-tech Park Wanda Plaza Co., Ltd.
项目地址: 大连市高新园区　**Location**: High-tech Zone, Dalian
室内设计: 北京清尚建筑设计研究院曾卫平商业设计
Interior Design: Beijing QingShang Landscape Architecture Design Institute　ZWP Commercial Design
主创设计: 曾卫平　**Chief Designer**: Zeng Weiping
建筑面积: 124 600平方米　**Building Area**: 124,600 m²
主要材料: GRG材质　**Major Materials**: GRG
采编: 康昭生　**Contributing Coordinator**: Kang Zhaosheng

网易家居网

大连高新万达广场的设计借助了形体之间材质、肌理、色彩、灯光及疏密关系上的穿插变化，使空间简约而不单调，热闹而不喧杂。

Home.163.com

Dalian High-tech Wanda Plaza utilizes alternate variation of material, texture, color, lighting and density relations to create a concise rather than drab and prosperous without clamorous space.

嘉宾点评
Honored Guest Comments

本案设计移步易景，各个空间设计手法迥异：室内空间的"模数化"设计，直街空间的平铺直叙，圆中庭的体块穿插，椭圆中厅的方格纹理，使空间达到了绚丽、精致、尊贵、高品位的艺术效果。

设计背景

区位特征

大连高新万达广场位于大连高新区旅顺南路北侧、七贤东路东侧，是万达集团在其发祥地大连打造的首个顶级商业综合体，也是万达集团2013年开业的两个A类店项目之一。

规划设计

大连高新万达广场总建筑面积约27万平方米，主要由商业综合体、室外商业街、写字楼、底商等组成，其中17万平方米为自持大商业，包括万达百货、万达影院、大歌星、大玩家、大型超市、大型电器商场等。

The project design has an effect of scenes changing with moving steps. Every space has its own design features: the interior space applying modular construction, the straight street space presenting plain and ungarnished impress, the roundish atrium interspersing blocks and the oval central nave adopting grid texture. The overall design reveals a gorgeous, delicate, noble and elegant art space.

Designing Backgrounds

Location

Dalian High-tech Park Wanda Plaza locates in Dalian High-tech Park with Lvshun South Road and Qixian East Road at its north and east respectively. Wanda Group's birthplace is in Dalian, where it built the first top level commercial complex which is one of the A Class projects opened in 2013.

Planning Design

The gross site area of Dalian High-tech Wanda Plaza is approximate 270,000 m². It is composed of commercial complex, outdoor pedestrian street, office building and underground commercial street, which including 170,000 m² self sustaining business, such as Wanda Department Store, Wanda Cinemas, Wanda Super KTV, Wanda Super Player, supermarket and super electronics store.

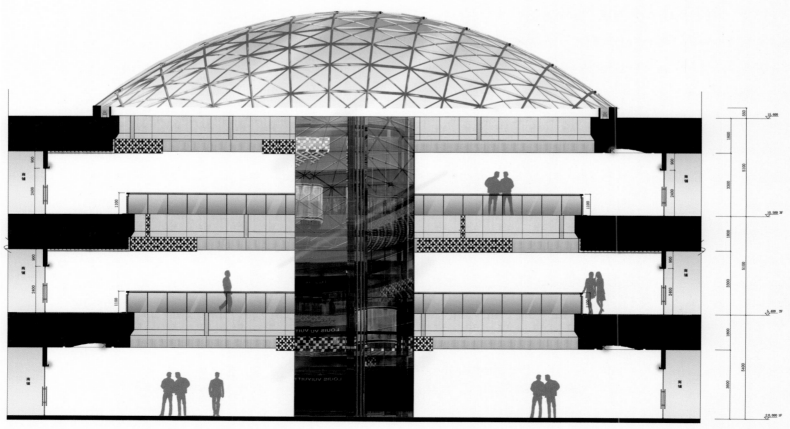

功能分区

商业中心一层以服装为主，二层以百货为主，三层以餐饮、娱乐为主，集购物、餐饮、文化、娱乐等多功能为一体，是万达集团在大连的顶级商业综合体。

室内空间："模数化"设计

室内空间中首次使用"模数化"设计，将外露的设备与顶面造型、侧立面、地面拼花等按照一定的比例模数有机排列，使长街达到井然有序、透视感强烈的视觉效果。

Functional Division

The commercial center is divided to costume zone on the ground floor, general merchandise on the 2nd floor and catering and entertainment on the 3rd floor, which is a top commercial complex in Dalian.

Interior Space: Modular Construction

The interior space uses modularization design for the first time. It is in accordance with certain proportion module organic arrangement to set up uncased equipment, ceiling modeling, side surface and parquet floor, bringing an orderly and perspective visual effect.

直街

在直街空间中，采用平铺直叙的表现手法，强调侧板的肌理和色彩的搭配，侧裙版GRG材质细微的凹凸和精致的型材收口等细节处理，令空间极具细节品质。

廊桥

在直街的中部廊桥里面，统一的排版模数，及材质、光色的对比，以及考究的炫金色金属质感烤漆玻璃，搭配内部透光效果，形成强烈的空间构成艺术效果，令廊桥成为直街的视觉焦点。

圆中庭

圆中庭采用了体块穿插的手法，两种颜色、质地强烈对比，直线体块在空间中的穿插，以及灯光营造的错位等，形成了交错有序的视觉效果，也使圆中庭的进深感加强，从而产生较大的视觉空间感。

侧立面

圆中庭侧立面分为两种材质与形态，分别采用灰金色和白色，同时注重细节处理，形成极具个性的中庭空间；观光电梯转折面采用透亮的方格元素，辅以背光处理，更显品质，令圆中庭空间时时流露出简约时尚、精致尊贵的艺术效果。

椭圆中庭

在椭圆中庭，围绕观光电梯周围的侧裙版采用方格纹理的发光玻璃，显露出晶莹剔透、美轮美奂的艺术效果；大面积侧裙版通过体块的穿插及漂浮设计手法，营造出轻盈灵动的空间特色；地面若隐若现的方格纹理拼花打破了地面的呆板沉闷，图案犹如花格地毯，尽显华贵典雅气质。

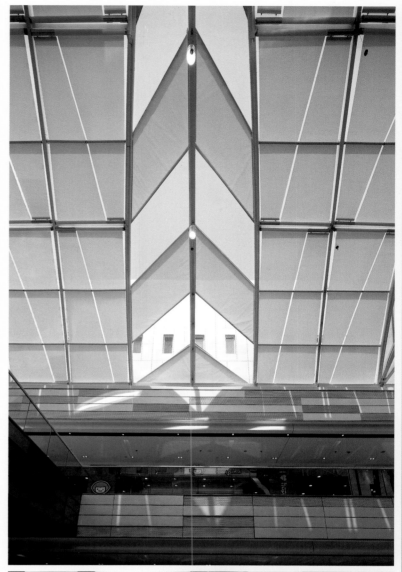

The Straight Street

In the straight street space, plain and ungarnished expression is witnessed. The texture and color collocation of side plate is emphasized. The GRE side skirt plate conducts particular treatment on subtle concave-convex and exquisite sectional material convergence, enhancing the detail quality.

Gallery Bridge

The middle part of the straight street sets a gallery bridge which facade has uniform layout module, texture and photochromic comparison and fine dazzled gold metallic paint glass, meanwhile goes with an inside lighting pervious effect, forming a strong space artistic flavor and making the gallery bridge to be a visual focus of the straight street.

Roundish Atrium

The roundish atrium has the layout of interspersing blocks with intense contrast of two colors and textures. Straight-line blocks interweave in the space plus dislocation lighting beaming, bringing staggered and orderly visual feeling, reinforcing the roundish atrium depth and outlining a larger visual space.

Side Surface

The side surface in the roundish atrium has two kinds of texture and types in gray gold and white respectively. The turning surface of sightseeing elevator adopts bright grid elements in backlighting processing. The design makes the atrium a concise, fashionable and noble space.

Oval Central Nave

In the oval central nave, the side skirt plate around sightseeing elevator employs luminescent glass in grid texture, presenting crystal clear and fabulous visual sense; massive side skirt plates utilizing interspersing blocks and floating design offer a lithe and dynamic space; the faintly discernible grid texture parquet, like a beautiful carpet, breaks surficial tediousness.

无立柱玻璃扶栏坡璃让视野更加开阔。纯白色人造石扶手拥有圆润细腻的质感，更具亲和力。
Glass balustrade without column makes a broader field of vision. Pure white artificial stone handrail brings soft and delicate touch.

GRG——侧立面使用的GRG是玻璃纤维加强石膏板，它是一种特殊改良纤维石膏装饰材料，造型的灵活性使其成为要求个性化的建筑师的首选，它独特的材料构成方式足以抵御外部环境造成的破损、变形和开裂。

GRG — The GRG material used in side surface is a glass fiber reinforced gypsum board. It is a special modified fiber gypsum decorative material. The arbitrary modeling feature makes it become the preferred material by personalized architects. GRG particular construction way makes it highly resist the external environment caused damage, deformation and cracking.

曾卫平 | Zeng Weiping

著名室内商业空间设计师、购物中心设计专家、教授级高级工程师。

Famous interior commercial space designer, shopping mall design expert and professor-level senior engineer.

所获荣誉 Awards & Honor

2008年 中国杰出中青年室内建筑师

2012年 第三届中国国际空间环境艺术设计大赛（筑巢奖）商业空间类金奖

2013年 中国环境艺术金奖

2013年 第四届中国国际空间环境艺术设计大赛（筑巢奖）商业展成空间方案类金奖

2014年 全国有成就的资深室内建筑师

2008 National Outstanding Indoor Architects of the Young and Middle Aged

2012 3rd China International Space Environment Art Design Competition (Nesting Award) and Commercial Space Gold Award

2013 China Environment Art Award, Gold Award

2013 4th China International Space Environment Art Design Competition (Nesting Award) Commercial Space Scheme Category Gold Award

2014 China Accomplished Senior Interior Architect

公司简介

曾卫平商业设计隶属于清华大学控股的北京清尚建筑设计研究院。近十年来，一直专注于商业空间领域的设计与研究，是国内领先的商业设计机构。其设计秉承"创新成就作品，细节决胜千里"的设计理念，通过在产业链上下游进行合理有效的延伸，根据客户的个性化需求，为各种商业项目提供咨询策划、建筑设计、室内设计、景观规划、平面设计等高品质全案服务。

Design Company Profile

Zeng Weiping Commercial Design Team is affiliated to Beijing QingShang Landscape Architecture Design Institute ZWP Commercial Design which strives to design and research in commercial space. They abide by the design concept of "innovation generating works, detail determining success". Through reasonable and effective extensions in upstream and downstream in the industry chain, and according to customer individual needs, they distribute targeted service for consultation, architecture design, interior design, landscape design and plane design, etc.

代表作品

武汉东湖汉街万达广场室内步行街、东莞东城万达广场室内步行街、杭州拱墅万达广场室内步行街、唐山爱琴海购物公园、成都五彩城。

Representative Works

Wanda Plaza Indoor Pedestrian Street in Donghu, Wanda Plaza Indoor Pedestrian Street in Dongguan, Wanda Plaza Indoor Pedestrian Street in Hangzhou, Tangshan Aegean Park, Chengdu Gorgeous City.

▼ 武汉东湖汉街万达广场室内步行街
Wanda Plaza Indoor Pedestrian Street in Donghu

▼ 东莞东城万达广场室内步行街
Wanda Plaza Indoor Pedestrian Street in Dongguan

▼ 杭州拱墅万达广场室内步行街
Wanda Plaza Indoor Pedestrian Street in Hangzhou

▼ 唐山爱琴海购物公园
Tangshan Aegean Park

▼ 成都五彩城
Chengdu Gorgeous City

后 记

第九届金盘奖自启动开始，受到了地产行业的广泛关注与支持，征集到了上千个项目的各类稿件，其中包括大量的实景图、效果图、设计图和设计说明等有关规划、建筑、景观等方方面面的资料，由于无法逐一现场考察参评项目，这些资料就是我们的嘉宾评委判断项目好坏与寻找项目特色的主要依据，在整个评选当中起到了至关重要的作用，在此感谢为金盘奖提供完整项目资料的供稿公司与设计师。

然而美中不足的是，很多来投稿的项目存在稿件质量不高，资料不齐，工程档案不完整甚至有错误等问题，给评委们做判断造成了困难，不得不说是整个评选工作的一大遗憾。一方面，一个好的项目需要恰当的展示，充实的项目资料才能让人对项目真正有所了解、欣赏，进而引发交流与探讨；另一方面，完整而详实的资料是项目得以在行业平台展示的前提，代表着参评方与推荐方的诚意。

最后，附上2015年久诺第十届金盘奖项目申报说明，相信在大家的信任与支持下，金盘奖主办方一定会在2015年再次打造出地产行业的高峰盛会。

2015年久诺第十届金盘奖项目申报说明

一、奖项定义
金盘奖是致力于为楼盘产品设计树立标杆，引领地产行业健康发展，为提升城市建筑面貌发挥积极作用的房地产设计类综合性大奖；是中国目前最具权威的民间地产设计大奖。

二、大奖口号
金盘奖力求搭建"中国好楼盘评选平台"，从"品质、艺术、人居、价值"的角度，表彰对社会大众来说真正优质的"好楼盘"项目。

三、评选宗旨
通过独立的评审机制，从专业、社会和文化层面，表彰具有突出社会意义、创新意识和人文关怀的优秀楼盘与空间作品。该奖项的目的还在于让生活中无所不在的建筑、空间得到公众应有的关注，推进社会进程。

四、基本原则
"权威、专业、公平、公正、公开"：以人文、艺术、创新等为标准，评论建筑设计艺术性、空间表现力、景观人文内涵及平面设计的适宜性和优异性。

五、奖项设置
金盘奖奖项包括年度最佳别墅、公寓、写字楼、酒店、综合楼盘、商业楼盘、旅游度假区、产业地产、样板房空间、售楼会所空间以及金盘年度人物。初赛分为华北、东北、西南西北、华中、华南、华东六大赛区，分赛区获奖项目将参加年度总评选。

六、组织机构
主办单位：《时代楼盘》杂志社
协办单位：中国建筑学会中国人居环境专业委员会
承办单位：金盘地产传媒有限公司
官方网站：金盘网www.KINPAN.com
媒体支持：搜房网、新浪乐居、搜狐焦点等

七、申报要求
（一）申报项目以2014年及其后竣工项目为主
（二）申报要求（获奖项目将刊登在《第十届金盘奖作品集》中）
　　A、总图：项目总平面图、区位图、交通分析图、鸟瞰图等
　　B、设计图：平面图、立面图、剖面图
　　C、实景图（不少于15张）：建筑、景观、室内、细节等
　　D、文字：工程档案信息、项目背景介绍（300字左右）、设计说明（800字左右）
　　E、主创设计师介绍：主创设计师姓名、照片、简介、代表项目名称等
　　F、VCR：主创设计师对项目亮点、理念的陈述，时长控制在2分钟以内
　　PS：• 图片需为电子文件，精度350dpi，TIF文件最小不低于10M
　　　　• 可登录金盘网自行下载申报表。
（三）格式要求
　　• 一级文件夹：项目名称
　　• 二级文件夹包括：总图文件夹；实景图文件夹；平、立、剖面图文件夹；文字文件夹；VCR文件夹
（四）截止时间：2015年7月7日

八、评奖流程
金盘奖采取公司申报与个人申报相结合的方式进行，最终由评委会评选产生。
（一）公司申报：申报公司根据奖项设置填好表格并提供项目资料。
（二）个人申报：设计师独立对其所有项目进行申报。
（三）初评：综合类奖项初评在六个分赛区进行，评委会对申报作品进行评议、投票、打分，选出分赛区获奖作品；
（四）网络投票：入围作品在金盘奖官方网站"金盘网"展示，并接受公众网络投票（每类奖项中得票最高的三个项目，均可在终评中多计一票）。
（五）评委考察：主办方组织评委对项目进行实地考察。
（六）终评：召开终评会议，投票产生获奖作品。

九、申报联系方式
登录金盘网www.kinpan.com 金盘奖专区获取第十届金盘奖评选申报表，根据项目类别仔细填写。
联系人：谢雪婷
联系电话：020-3206 9300-1017
联系邮箱：shidailoupan@kinpan.com
联系地址：广州市萝岗区科学大道中99号科汇一街7号3层
邮编：510 000
传真：020-3206 9330

十、附则
1、申报项目必须为本人或本公司原创作，具有完整版权和使用权，项目中涉及肖像权、著作权、知识产权等法律事宜均由参赛者本人承担相关责任；
2、为了保证第十届金盘奖评选的公平、公正、客观以及其专业性，参赛者必须保证参赛资料真实、准确。主办方将对参赛材料进行形式审查，不完整、不正确或不符合参赛要求的参赛材料，举办方有权取消参评者的参赛资格。

Postscript

The 9th Kinpan Award has received extensive attention and support from real estate industry since it was launched. It has collected hundreds contributions of various types, including a large amount data on planning, architecture, landscape, such as realistic pictures, renderings, design drawings and design descriptions. Since we could not arrange a site inspection for every project, the data holds grounds for us to evaluate, therefore, they play a significant role in the process of the evaluation. Here we express our sincere thanks to those companies and designers who provide complete project data.

However, there are still some projects with low quality, incomplete data and project credits, or even incorrect data, resulting difficulties in evaluation. It is a pity for the evaluation and selection. On one hand, a good project needs proper display with substantial information and data to make it be known and appreciated so as to stimulate deeper communication and discussion; on the other hand, complete and detailed data is a premise for a project to be shown on an industry platform, since they represent the sincerity of contributors and recommenders.

Last but not least, the 2015 Jiunuo 10th Kinpan Award Project Application Description is attached. Bearing everyone's trust and support, we will deliver a premier real estate event in 2015.

2015 Jiunuo 10th Kinpan Award Project Application Description

1. Definition
The Kinpan Award is a comprehensive award of property design, which is committed to establish a benchmark for property design, guide a healthy development of real estate industry and play a positive role in enhancing urban building image. Currently, it is the most authoritative nongovernmental award of property design in China.

2. Slogan
The Kinpan Award strives to set up an "evaluation platform of good properties in China" and commend genuine "good properties" for the public from aspects of "quality, art, habitat and value".

3. Objective
Through independent review mechanism on professional, social and cultural aspects to boost preeminent properties and space works possessing outstanding social meaning, innovation consciousness and humanistic care. The Kinpan Award aims to let properties and spaces existing in our life harvest deserved attention so as to push forward social development.

4. Principles
The Kinpan Award takes "authority, profession, fairness, justice, open" as evaluation principles and culture, art, innovation as core value standards. The award reviews the suitability and virtue of architecture designs from views of art, space, humanistic connotation of landscape and plan design.

5. Awards
The Kinpan Award is composed of the Best Villa, the Best Apartment, the Best Office Building, the Best Hotel, the Best Comprehensive Property, the Best Commercial Property, the Best Tourist Resort, the Best Industrial Property, the Best Show Flat Space, the Best Sales Center Space and Annual Kinpan Character. The preliminary contests are in 6 competition areas: Northern China, Northeast China, Southwest and Northwest China, Central China, Southern China and Eastern China.

6. Organizations
Sponsor: Times House Magazine
Co-organizer: Human Settlement Specialized Committee of Architectural Society of China
Organizer: Kinpan Estate Media Ltd.
Official Website: www.kinpan.com
Media Support: SouFun.com, house.sina.cn, focus.cn, etc.

7. Project Requirement
7.1 Project completed in 2014 and thereafter
7.2 Data Requirement (Award-winning projects are going to be published in the 10th Kinpan Award Files)
7.2.1 General drawings: master plan, location map, traffic analysis, aerial view, etc.
7.2.2 Design drawings: plan, elevation, section, etc.
7.2.3 Photos (more than 15 sheets): building, landscape, interior, detail, etc.
7.2.4 Text: project credit, background (about 300 words), design description(about 800 words)
7.2.5 Chief designer introduction: name, photo, profile, masterpiece, etc.
7.2.6 VCR: Chief architect's interpretation for highlights and concepts of the project, within 2 minutes.
PS. Photos and drawings should be electronic documents of 350dpi.
   TIF format document should be no less than 10M.
   Application form is available on www. Kinpan.com

7.3 Submit format
First Level Folder: Project's Name
Second Level Folder: General Drawing Folder, Realistic Pictures Folder, Plan, Elevation and Section Folder, Text Folder, VCR Folder

7.4 Deadline
July 7th, 2015

8. Evaluation Procedure
The Kinpan Award accepts both company participant and individual participant, and the award-winning projects are reviewed and elected by judging panel.
8.1 Company participant: fills out the form according to the awards setting and offers project data.
8.2 Individual participant: designers declare individual original projects.
8.3 Preliminary evaluation: comprehensive awards will be evaluated in 6 divisions respectively, and judging panel review and select through appraisal, vote, and score procedure.
8.4 Internet voting: shortlisted projects will be displayed on www.kinpan.com and decided by public voting (the three projects with most votes of each award will gain a particular point in the final evaluation).
8.5 Judge inspection: the sponsor arranges judge site inspection.
8.6 Final evaluation: the sponsor convenes final evaluation and votes winning projects.

9. Contact Information
Log in www.kinpan.com to access application form of the 10th Kinpan Award and fill out according to corresponding project types.
Contact: Xie Xueting
Hotline: 020-3206 9300-1017
E-mail: shidailoupan@kinpan.com
Address: 3rd floor, No.7 1st Street, International Creative Valley, No.99 Science Avenue Middle, Luogang District, Guangzhou
Postcode: 510000
Fax: 020-3206 9330

10. Supplementary Articles
10.1 Project should be original works with complete copyright and right of use. The legal matters concerning portraiture right, copyright and intellectual property of the works should be assumed by contributors.
10.2 To guarantee the fairness, justice, objective and profession of the forthcoming 10th Kinpan Award, participants are asked for submitting true and accurate project information and data which will be investigated by the host who has the right to cancel the qualification for participating in competition if the relevant information is incomplete, inaccurate or below entry requirements.

图书在版编目（CIP）数据

第九届金盘奖获奖作品集．空间类/广州市唐艺文化传播有限公司编著．－－北京：中国林业出版社，2015.3

ISBN 978-7-5038-7905-0

Ⅰ．①第… Ⅱ．①广… Ⅲ．①建筑设计－中国－图集②室内装饰设计－中国－图集 Ⅳ．① TU206 ② TU238

中国版本图书馆CIP数据核字（2015）第043241号

第九届金盘奖获奖作品集　空间类

编　　　著	广州市唐艺文化传播有限公司
责 任 编 辑	纪　亮　王思源
策 划 编 辑	段妮静
流 程 编 辑	黄　姗
文 字 编 辑	张　芳
英 文 编 辑	冯亭亭
装 帧 设 计	陈阳柳

出 版 发 行	中国林业出版社
出版社地址	北京西城区德内大街刘海胡同7号，邮编：100009
出版社网址	http://lycb.forestry.gov.cn/
经　　　销	全国新华书店
印　　　刷	深圳汇亿丰印刷科技有限公司
开　　　本	245 mm×325 mm
印　　　张	15.5
版　　　次	2015年3月第1版
印　　　次	2015年3月第1次印刷
标 准 书 号	ISBN 978-7-5038-7905-0
定　　　价	268.00元（USD 48.00）（精）

图书如有印装质量问题，可随时向印刷厂调换（电话：0755-82413509）。